The Saber and Scroll Journal

Also from Westphalia Press
westphaliapress.org

The Idea of the Digital University

Dialogue in the Roman-Greco World

The History of Photography

International or Local Ownership?: Security Sector Development in Post-Independent Kosovo

Lankes, His Woodcut Bookplates

Opportunity and Horatio Alger

The Role of Theory in Policy Analysis

The Little Confectioner

Non-Profit Organizations and Disaster

The Idea of Neoliberalism: The Emperor Has Threadbare Contemporary Clothes

Social Satire and the Modern Novel

Ukraine vs. Russia: Revolution, Democracy and War: Selected Articles and Blogs, 2010-2016

James Martineau and Rebuilding Theology

A Strategy for Implementing the Reconciliation Process

Issues in Maritime Cyber Security

A Different Dimension: Reflections on the History of Transpersonal Thought

Iran: Who Is Really In Charge?

Contracting, Logistics, Reverse Logistics: The Project, Program and Portfolio Approach

Unworkable Conservatism: Small Government, Freemarkets, and Impracticality

Springfield: The Novel

Lariats and Lassos

Ongoing Issues in Georgian Policy and Public Administration

Growing Inequality: Bridging Complex Systems, Population Health and Health Disparities

Designing, Adapting, Strategizing in Online Education

Pacific Hurtgen: The American Army in Northern Luzon, 1945

Natural Gas as an Instrument of Russian State Power

New Frontiers in Criminology

Feeding the Global South

Beijing Express: How to Understand New China

The Rise of the Book Plate: An Exemplative of the Art

The Saber and Scroll Journal

Volume 12, Number 1 • Summer 2024

Gilda Bour, Editor-in-Chief

Westphalia Press
An imprint of Policy Studies Organization

The Saber and Scroll Journal
Volume 12, Number 1 • Summer 2024

Westphalia Press
An imprint of Policy Studies Organization
1367 Connecticut Ave NW
Washington, D.C. 20036
info@ipsonet.org

ISBN: 978-1-63723-408-2

Cover and interior design by Jeffrey Barnes
jbarnesbook.design

Daniel Gutierrez-Sandoval, Executive Director
PSO and Westphalia Press

Updated material and comments on this edition
can be found at the Westphalia Press website:
www.westphaliapress.org

APUS

The Saber and Scroll Journal
Volume 12, Number 1 • Summer 2024

Historical Studies Honor Society

PSO

On the cover

In this art miniature from an English psalter, a knight kneels before his horse as he prepares to set off on a crusade. The crosses on the surcoat reflected Christian moral responsibility and signaled the wearers' willingness to fight and risk their lives for God. Knights and noblemen would take vows that, while religious, were also militant.

Image Credit: A depiction of the Knight with Horse from the 'Westminster Psalter': Royal 2 A XXII, Part 1 ff. 1-218, Psalter, with litany, prayers, and Easter tables. England, S. (Westminster), c.1200-1250, Latin.

Welcome Letter

Welcome to the 2024 Summer Issue of *The Saber and Scroll Journal.* This issue includes articles covering a wide range of historical topics and periods. We travel to South America with an article examining and interpreting leftwing insurgencies in Colombia. In this article, the author takes us from the 1960s to now, in a study of leftwing rebel groups such as FARC, ELN, and M19. We go back much farther in history with the Crusader vows article, to discover how politicization changed the vows during and after the Crusading Period. We move back to the 20th century with two articles, one that places us in the tumultuous progressive period with an article about what led to the *Moore v. Dempsey* Supreme Court case and another that explores the doctors who developed chemotherapy drugs that advanced the study of cancer research and treatments. A museum review by author Aisha Manus provides a behind-the-scenes tour of a museum in Georgia. An exhibit review takes readers over the pond to explore a museum exhibit about the last witch trials in Northern Ireland.

Book reviews are vital portions of historical review and are not always easy to write. While there is a desire to provide a positive review, a critical assessment of the book must occur to provide honest feedback to prospective readers and help place the work within a proper historical lens. Book reviews provide context to future readers and a comparison to other works in the field. This edition of the journal provides two essential book reviews, which provide readers valuable insight into the content of each book, *The Luzon Campaign 1945: MacArthur Returns and Who's Who in the Age of Alexander and his Successors.*

I want to acknowledge Jeffrey Ballard for all his hard work and helping usher me through the editor role this past year. He leaves me big shoes to fill as I step into the role he is entrusting me with editor-in-chief. I hope to continue the tradition of bringing historical literacy to the forefront and bringing history alive, as he has done.

In closing, I want to thank the authors, contributors, readers, peer reviewers, and the editing team who put in a lot of work to make each issue happen. The journal is a labor of love, and each issue reflects this effort. We hope you, the readers, find value in this issue's articles and reviews.

Gilda Bour
Editor-in-Chief
The Saber and Scroll Journal
July 23, 2024

doi: 10.18278/sshj.12.1.1

Carta de bienvenida

Bienvenidos a la edición de verano de 2024 de *The Saber and Scroll Journal*. Esta edición incluye artículos que cubren una amplia gama de temas y períodos históricos. Viajamos a Sudamérica con un artículo que examina e interpreta las insurgencias de izquierda en Colombia. En este artículo, la autora nos lleva desde la década de 1960 hasta ahora, en un estudio de grupos rebeldes de izquierda como las FARC, el ELN y el M19. Nos remontamos mucho más atrás en la historia con el artículo sobre los votos de las Cruzadas, para descubrir cómo la politización cambió los votos durante y después del Período de las Cruzadas. Nos trasladamos al siglo XX con dos artículos, uno que nos sitúa en el tumultuoso período progresista con un artículo sobre lo que condujo al caso *Moore v. Dempsey* de la Corte Suprema y otro que explora a los médicos que desarrollaron medicamentos de quimioterapia que hicieron avanzar el estudio de la investigación y los tratamientos del cáncer. Una reseña de un museo a cargo de la autora Aisha Manus ofrece un recorrido tras bastidores por un museo en Georgia. Una reseña de una exhibición lleva a los lectores al otro lado del charco para explorar una exhibición de un museo sobre los últimos juicios de brujas en Irlanda del Norte.

Las reseñas de libros son partes vitales de la revisión histórica y no siempre son fáciles de escribir. Si bien existe el deseo de proporcionar una reseña positiva, se debe realizar una evaluación crítica del libro para proporcionar comentarios honestos a los posibles lectores y ayudar a ubicar el trabajo dentro de una perspectiva histórica adecuada. Las reseñas de libros brindan contexto a los futuros lectores y una comparación con otras obras en el campo. Esta edición de la revista ofrece dos reseñas de libros esenciales, que brindan a los lectores una valiosa perspectiva del contenido de cada libro, *The Luzon Campaign 1945: MacArthur Returns* y *Who's Who in the Age of Alexander and his Successors.*

Quiero agradecer a Jeffrey Ballard por todo su arduo trabajo y por ayudarme a desempeñar el papel de editor durante el año pasado. Me deja un gran desafío al asumir el papel que me ha encomendado como editor jefe. Espero continuar con la tradición de poner en primer plano la alfabetización histórica y dar vida a la historia, como él lo ha hecho.

Para finalizar, quiero agradecer a los autores, colaboradores, lectores, revisores y al equipo editorial que han dedicado mucho trabajo para que cada número se publique. La revista es un trabajo hecho con amor y cada número refleja este esfuerzo. Esperamos que ustedes, los lectores, encuentren valor en los artículos y reseñas de este número.

Gilda Bour
Editora principal
The Saber and Scroll Journal
23 de julio de 2024

编者按

欢迎阅读《军刀与卷轴杂志》2024年夏季刊。本期杂志收录了涵盖广泛历史主题和时期的文章。第一篇文章聚焦南美洲，检验并诠释了哥伦比亚的左翼叛乱。在这篇文章中，作者带我们研究了从1960年代到现在的左翼叛乱组织，例如哥伦比亚革命武装力量(FARC)、民族解放军(ELN)和“四一九运动”武装部队(M19)等。我们通过一篇关于十字军誓言的文章追溯更远的历史，以发现政治化如何在十字军东征期间和之后改变了誓言。之后的两篇文章回到20世纪，其中一篇将我们带入动荡的进步时期（关于是什么导致摩尔诉邓普西案这一最高法院案件），另一篇则探究了开发化疗药物的医生，这些药物推动了癌症研究和治疗研究。作者Aisha Manus的博物馆评论文提供了一所位于佐治亚州的博物馆的幕后之旅。一篇展览评论带领读者跨越大西洋去探索有关北爱尔兰最后一次女巫审判的博物馆展览。

书评是历史评论的重要组成部分，写起来并不容易。虽然我们希望提供正面的评论，但必须对书进行批判性评估，以便向潜在读者提供诚实的反馈，并帮助将作品置于适当的历史视角中。书评为未来的读者提供了情境，并将书评对象与该领域的其他作品进行了比较。本期杂志提供了两篇重要的书评，为读者提供了对两本书内容的宝贵见解，这两本书分别为《1945年吕宋岛战役：麦克阿瑟归来》(The Luzon Campaign 1945: MacArthur Returns)和《亚历山大及其继任者时代的名人录》(Who’s Who in the Age of Alexander and his Successors)。

我要感谢Jeffrey Ballard在过去一年的辛勤工作，他帮助我完成了编辑工作。当我接任他委托给我的主编一职时，他留给我的是巨大的压力。我希望继续延续这一传统，将历史素养放在首位，让历史鲜活起来，就像他所做的那样。

最后，我要感谢作者、投稿人、读者、同行评审员和编辑团队，他们为每一期杂志的出版付出了辛勤努力。本刊是一份充满爱的劳动，每一期杂志都体现了这种努力。我们希望读者能从本期的文章和评论文中找到价值。

Gilda Bour
执行编辑
《军刀与卷轴杂志》
2024年7月23日

Museum Employees Only: Inside The Knox Foundation Center for the Preservation and Study of CSRA History

Aisha Manus

Independent Historian

The main entrance to the Augusta Museum of History. Built in the 1990s as part of a revival effort for the Riverwalk area, part of the building design incorporated the use of older, historical pieces from other buildings within the CSRA, such as the columns from the Georgia Railroad and Banking Company. [The Augusta Museum of History. Photo by Aisha Manus.]

The experience the average person has when they visit a museum, whether it is a large, national, or state-level museum or a small, hometown museum, is a curated experience crafted by the hard work of many people in the background. The average person seldom sees what goes on behind closed doors to make their museum experience possible, but the Augusta Museum of History, located at 560 Reynolds Street in Augusta, GA is changing that. Each September, on the first three Saturdays of the month,

doi: 10.18278/sshj.12.1.2

they offer museumgoers a chance to go behind the locked doors of The Knox Foundation Center for the Preservation and Study of CSRA History and the Educational Resource Center, where they can learn about the work that not only goes in preserving each item the museums receives but to see things that are generally not on display.

Headed by Alena Rensch, the museums registrar, and Dana Wilde, the Wilkinsons Collections intern, the tour, which consisted of myself, a grandmother with her two grandchildren, and a couple that just happened to come into the museum as the tour was starting, found us walking through the Transportation Corridor of the museum, past the massive 1914 locomotive, a 1920s trolley, and a 1930s gas station on display, through a door marked for employees only and into a small lobby for the entrance of the Knox Center. Built in 2003, this 10,000 square-foot addition houses 93 percent of the museums collection, nearly all of which comes from private donors. It also houses a research library, archives, and an artifact conservation and preparation laboratory. The latter of the two was our first stop.

So as not to contaminate nor accidentally ruin an item in the lab, food or drinks were strictly forbidden and we were asked to leave our bags on a cart just outside of the room. There were three rows of chairs in front of a long table, where recently donated items sat waiting for Ms. Rensch to process as part of her job as registrar. She walked us through the process of assessing the age and material of an item so it could be decided how to care for the item not just then but also in the future. The processes that she explained were very in-depth, but the language used was very friendly and simplistic enough for even the youngest visitors to understand. One of the examples she demonstrated was using a piece of tissue paper in-between each sheet of newspaper to balance out the pH, as the paper is considered very acidic.

The sign designating the entrance to The Knox Foundation Center for the Preservation and Study of CSRA History. Though we were inside a lobby, the exposed brick walls indicated we were entering an entirely new building. [The Augusta Museum of History. Photo by Aisha Manus.]

Next she taught us about the different agents of deterioration, how each can damage items overtime, and how the museum combats each of these agents. To deal with the most common

Above: Augusta Museum of History registrar Alena Rensch showing us the process of placing tissue paper between sheets of newspaper to balance out the pH and combat the agent of deterioration that is acidity. [The Augusta Museum of History. Photo by Aisha Manus.]

Right: Dana Wilde, the Wilkinsons Collections intern, delicately turns the page of a letter from a solider stationed in an undisclosed location in France during WWI. Dated November 17th, 1917, the letter is addressed to a Mr. J. Davenport and comes from a 1st Lt. Robert B. Durham. The letter reads quite jovial as Lt. Durham has clearly just arrived in France, having spent a week in Paris, which was full of "swell women. My, they are plentiful." The second letter from the now Capt. Durham in Bordeaux to Mr. Davenport lay adjacent to the first. This one, dated April 10th, 1919, is much more serious in tone, with Capt. Durham relaying that he spent 10 months in the trenches being gassed and 50,000 men were either killed or wounded in his division. The juxtaposition of these two letters was a wonderful way to show how war can change a person so quickly. [The Augusta Museum of History. Photo by Aisha Manus.]

agent, temperature and humidity, Ms. Rensch explains that the museum is kept at 70 degrees Fahrenheit and 50 percent humidity year-round, as that is the most optimal temperature and humidity for most items in the collection. She also explained the effects of light and pests as agents of deterioration, as well as the oils on our own hands, have on items. She stressed the use of gloves when handling objects, but not just as a protection for the item but for us as well, as she explained that sometimes older items were made with chemicals or processes that we know to be toxic today, for example, old leather books.

After answering our questions about processing donations and the care and maintenance of those items, we moved into my favorite area, the Research Library. While the room was small, an assortment of books related to the local history of the Central Savannah River Area (CSRA) filled the bookshelves. Priceless portraits of people and places of importance hung on the walls. In the center of the room there was a unique display of the more common artifacts that the average person has in their home, such as letters, photographs, and clothes. I naturally gravitated to the set of letters from a soldier stationed in France during World War I. Reading these letters, sent 17 months apart, you can see how his tone changed, and how war had changed him. When I wanted to read the second page, Dana delicately separated the pages with a small spatula. She also made sure to do just one letter at a time, as she informed me that for the sake of preservation items cannot touch one another if at all possible.

Two other items that interested me on the table were an early turn of the 20th century era apology letter that was super funny in that it was a check-the-box style and an old scrapbook. The letter was a lovely reminder that the people of the past were people just like us with joy in their hearts and laughs in their bellies who also needed to sometimes apologize for their behavior at the friends party. The scrapbook, which was full of articles and a local family's history, was a reminder that everyone is or can be their family's personal historian. This particular scrapbook was actually damaged by pipe bursting in the storage facility in January 2023 which led to a great discussion with both Ms. Rensch and Ms. Wilde about the restoration process, of using tissue paper between each page to absorb the excess water. They had to apply this process for multiple weeks to save the several boxes subjected to the leak. This also led to a discussion about the museums ongoing digitizing project. While new items are digitized during their initial processing, they are still working tirelessly to digitize items already in their collection, relying on the time and efforts of their interns, like Ms. Wilde who is currently working on digitizing a collection of photos. The museums hope is to have a public online collection that is cloud-based so it can be accessed anywhere but that could be years from now.

The price of the tour is just the cost of admission, a mere $5 for adult non-members, dropping down to $4 for seniors and $3 for children over 6. Children age 6 and under are free. The museum is open Thursday through Saturday

Alena Rensch presenting us with a tongue and cheek apology card from the turn of the 20th century. The card was an apology letter using a check the box style. The boxes you could check had to do with being too drunk, or too loud, or other funny reasons for causing a scene at the party the previous night. Unfortunately, this card was unused, so there was no fun story we could have imagined from the person who either gave or received this card. But it still says a lot about the person who would have owned the card to use one day. [The Augusta Museum of History. Photo by Aisha Manus.]

from 10am to 5pm and on Sunday from 1pm to 5pm. The private tour is only offered on select Saturdays in September at set times, but guests are able to explore the museum as well before or after. The Augusta Museum is one of only 20 museums in the state of Georgia that is accredited by the American Alliance of Museum, and this wonderful behind the scenes tour was proof of the adherence to the highest professional standards that got them that coveted seal of approval. I encourage others to make a trip to the museum next September for a look behind the doors and into Augustas deep and rich history, and the processes that goes into uncovering it. For more information, check out their website at www.augustamuseum.org or call 706-722-8454.

Ideological Warfare: Interpreting Leftwing Insurgencies In Colombia

Valbona Karanxha EdD

Abstract

Since the early 1960s, Colombia has been home to violent leftwing insurgencies, some of which are still active. The corollary for the rise of the guerrillas was the deadlock in domestic politics and a dismal agrarian law since the 1930s. The disagreements between the liberals and conservatives at the time led to ideological clashes and the founding of leftwing rebel groups such as FARC, ELN, and M19. On the other side of the political spectrum, the conservatives relied on the United Self-Defenses of Colombia (AUC)—paramilitary units—to counterbalance the rise of leftwing insurgencies. The state, as the leading actor in stabilizing the country, faced rising political violence spurred by ideological intolerance. More importantly, the state lacked a trained military and adequate police force to combat the guerrillas, the drug cartels and stop the conflict from spreading.

This research uses FARC and other leftwing uprisings in Colombia to create a framework for ideological insurgencies, focusing primarily on their evolution and disintegration once the ideology faded. The research found that there is a positive correlation between ideology and insurgency. The insurgency's success, such as the number of members it enlists, the territories it captures, the population support, and financial security, depended on the strength of its ideological base, the *Zeitgeist*, and the international support. At the same time, when these factors weakened, then ideology became irrelevant.

The study revealed that grievances and politics of contentions drove the rhetoric with the intent to create a propagandistic machine to enlist guerrillas and secure a solid financial basis to leverage the cause and achieve political goals. However, once the ideology became outdated and failed, the insurgency changed its course and turned into a greed-based insurgency seeking to maximize profits through illicit activities.

Keywords: Insurgency, grievance-based insurgency, greed-based insurgency, ideology, leftwing ideology, Marxism, agrarian revolution, rebelocracy, multinational operations, military operations, Plan Colombia, military assistance, guerrilla warfare, counterinsurgency

doi: 10.18278/sshj.12.1.3

Guerra ideológica: interpretar las insurgencias de izquierda en Colombia

Resumen

Desde principios de la década de 1960, Colombia ha sido hogar de violentas insurgencias de izquierda, algunas de las cuales todavía están activas. El corolario del ascenso de las guerrillas fue el estancamiento de la política interna y una deprimente ley agraria desde los años treinta. Los desacuerdos entre liberales y conservadores en ese momento llevaron a enfrentamientos ideológicos y a la fundación de grupos rebeldes de izquierda como las FARC, el ELN y el M19. En el otro lado del espectro político, los conservadores confiaron en las Autodefensas Unidas de Colombia (AUC), unidades paramilitares, para contrarrestar el aumento de las insurgencias de izquierda. El Estado, como actor principal en la estabilización del país, enfrentó una creciente violencia política estimulada por la intolerancia ideológica. Más importante aún, el Estado carecía de un ejército capacitado y de una fuerza policial adecuada para combatir a las guerrillas, los cárteles de la droga y evitar que el conflicto se extendiera.

Esta investigación utiliza a las FARC y otros levantamientos de izquierda en Colombia para crear un marco para las insurgencias ideológicas, centrándose en su evolución y desintegración una vez que la ideología se desvanece. La investigación encontró que existe una correlación positiva entre ideología e insurgencia. El éxito de la insurgencia, como el número de miembros que recluta, los territorios que captura, el apoyo de la población y la seguridad financiera, dependieron de la fuerza de su base ideológica, el *zeitgeist*, y del apoyo internacional. Al mismo tiempo, cuando estos factores se debilitaron, la ideología se volvió irrelevante.

El estudio reveló que los agravios y la política de contiendas impulsaron la retórica con la intención de crear una máquina propagandística para reclutar guerrilleros y asegurar una base financiera sólida para aprovechar la causa y lograr objetivos políticos. Sin embargo, una vez que la ideología quedó obsoleta y fracasó, la insurgencia cambió de rumbo y se convirtió en una insurgencia basada en la codicia que buscaba maximizar las ganancias a través de actividades ilícitas.

Palabras clave: Insurgencia, insurgencia basada en agravios, insurgencia basada en la codicia, ideología, ideología de izquierda, marxismo, revolución agraria, rebelocracia, operaciones multinacionales, operaciones militares, Plan Colombia, asistencia militar, guerra de guerrillas, contrainsurgencia

意识形态战：解析哥伦比亚的左翼叛乱

摘要

自20世纪60年代初以来，哥伦比亚一直是暴力左翼叛乱的发源地，其中一些叛乱仍然活跃。游击队崛起的必然结果是国内政治的僵局和自20世纪30年代以来惨淡的土地法。当时，自由派和保守派之间的分歧导致了意识形态冲突，并导致了哥伦比亚革命武装力量、民族解放军和M19等左翼叛乱组织的成立。在政治光谱的另一边，保守派依靠哥伦比亚联合自卫队(AUC)——准军事部队——来制衡左翼叛乱的崛起。作为稳定国家的主要行动者，哥伦比亚面临着意识形态不宽容引发的不断上升的政治暴力。更重要的是，国家缺乏训练有素的军队和足够的警察部队来打击游击队、贩毒集团并阻止冲突蔓延。

本研究利用哥伦比亚革命武装力量和哥伦比亚的其他左翼起义来创建一个意识形态叛乱框架，聚焦于意识形态消退后叛乱的演变和瓦解。研究发现，意识形态与叛乱之间存在正相关关系。叛乱分子的成功，例如其招募的成员数量、占领的领土、人口支持和财政安全，取决于其意识形态基础(Zeitgeist)的强度和国际支持。与此同时，当这些因素减弱时，意识形态就变得无关紧要了。

研究显示，不满情绪和政治争论推动了这些言论，其目的是创建一个宣传机器来招募游击队，并确保坚实的财政基础来利用这一事业并实现政治目标。然而，一旦意识形态变得过时并失败，叛乱分子就改变了路线，变成了以贪婪为基础的叛乱组织，寻求通过非法活动来实现利益最大化。

关键词：叛乱，基于不满的叛乱，基于贪婪的叛乱，意识形态，左翼意识形态，马克思主义，土地革命，叛乱(rebelocracy)，多国操作，军事操作，哥伦比亚计划，军事援助，游击战，反叛乱

In the last thirty years, Colombia has been at the center of the U.S. policy for Latin America due to the ongoing armed conflict between the state and leftwing insurgents. Decades of counterinsurgency (COIN) proved challenging to deter the most prolonged political and military insurgency from controlling substantial parts of rural Colombia.[1] Throughout the 1980s, the country's future looked grim. The proliferation of drug production and trafficking added another layer to the conflict. The Revolutionary Armed Forces of Colombia (FARC) controlled strategic infrastructure and governed their territories inside Colombia's national territory. They had uncontested power and ran a shadow "government" parallel with the state in the disputed zones. The inside conflict threatened the state's legitimacy. Colombia risked losing territorial sovereignty to the insurgents in the zones where the state's presence was weak or challenged. As the conflict prolonged for decades, the Colombian state had exhausted a great deal of resources to fight the rising leftwing insurgency. Still, more importantly, it lacked a trained military, technology, and adequate strategies to combat the rebels.[2]

FARC emerged into the conflict as an extremist faction of the Colombian Communist Party (PCC Partido Comunista Colombiano) in the 1960s. Like many ideological insurgencies of their time, such as the Vietcong, the Cuban Revolution, and Cuban-instigated insurgencies in Latin America, it adopted the same zeitgeist rhetoric.[3] The leftwing ideology served as the bond to keep the organization together, seeking to cause an agrarian revolution with the potential to seize power and establish a regime.

The role ideology plays in political violence can be of great interest to understanding the political game and how political actors behave. Insurgencies and political movements tend to degrade when ideology is no longer relevant to the cause. Once the doctrine fails, the movement morphs from a grievance-based to a greed-based insurgency, initially in efforts to fund the war until it turns profitable. This study will try to explain the following dilemmas while analyzing the conflict's particularities.

- How does an insurgent group use ideology to justify political violence?
- How do internal and external factors affect ideological insurgency's evolution and dissolution?

There is a plethora of literature on ideology and insurgency. However, analyzing the transition of FARC from grievance-based to greed-based insurgency needs a more critical approach. FARC has a place in the literature about rebellions and rebelocracy, yet the role of the ideology as the bonding agent is less evident. The studies concerning FARC focus mainly on military operations and capacities as part of the conflict that spurred violence. Other works introduce a new phenomenon in discussing civil wars: Wartime and social order. Such a study uses the case of Colombia to postulate that insurgencies tend to create systems within the chaos to perpetuate their existence. It analyses how social order functions in conflict zones. While the argument is valid and sound, it focuses primarily on FARC as an illegitimate organization able to compete with the state. Yet, it leaves out ideology and propaganda as the moving force in the conflict.

This research seeks to create a framework for ideological insurgencies, focusing primarily on their evolution as part of a larger conflict and the disintegration process once the ideology weakens.

Revolutionary Armed Forces of Colombia (FARC) insurgents. [Institute for National Strategic Studies (INSS.) - Strategic Assessment 1998: Engaging Power for Peace, Chapter 7, The Americas.]

Roots of the Colombian Insurgency: Ideology and Propaganda

Insurgency thrives on ideology and propaganda.[4] It is a bonding agent between the cause and the population affected by the injustice. The basis of the Colombian insurgency stemmed from a revised form of Marxist ideology combined with lessons and practices from the Chinese agrarian revolution.[5] In the early 1900s, a new wave of political movements fueled by Marxist ideology spread around the globe. The events gave birth to communist systems across Europe and globally. In 1930, Colombia founded the communist party with the help of the Soviet Union and the Comintern.[6]

Hence, the catalyst for the rise of the leftwing insurgencies in Colombia were internal failures in land policy, the agrarian reforms, and finding an equilibrium between the agro-economic sector and small landowners. The discontent with the agrarian laws and property rights prompted a conflict and the founding of armed rebel groups in the 1960s seeking more than land and wealth distribution.[7] The Marxist concepts of class warfare and primitive accumulation served as the ideological basis of the Colombian insurgency since its inception—and the revolution was the only solution to the problem. The agrarian revolution found a solid basis in the communist ideology and propaganda due to the country's socio-economic situation coupled with the unsolved land problems it had inherited from the past.

Propaganda is a powerful instrument to generate support, whether local or international. The key to an intense propaganda campaign is the heated rhetoric and the diffusion of firm beliefs that the war is justified and it is the only means to appease the repertoire of contentions.[8] The end objective of ideological campaigns is always fundraising, recruiting, collecting taxes locally, and finding outside support from state or non-state actors. Effective propaganda during the conflict helps legitimize the insurgency's goals, discredit the government, and internationalize the armed struggle by bringing a movement's message to a broader audience.[9] The international leftist movements, new political models, and propaganda campaigns against the state made Colombian insurgencies unique for the time and the entire region. Combating the insurgency was a paramount task. It required understanding how and where it operated and understanding the power of ideology used to recruit new members. It meant knowing the enemy of the state. The conflict had many particularities attributed to Colombia's historical past and the ongoing political deadlock.[10]

Insurgency and Ideology

Nowadays, a general perception of ideology at the center of political violence is that it lacks relevance. Can ideological movements fall into the category of social movements, and as such, can they convert into armed insurgency? Marxist insurgencies have fallen out of the political discourse after the collapse of communism.[11] A Marxist insurgency strives to establish institutions that seek some form of an egalitarian society, a planned economy, and a highly centralized government. As a transnational ideology, Marxism seeks support from various countries or individuals who embrace the cause.

Insurgency is a prolonged political-military activity aiming to control a country's territory and resources through non-statal military force and illegal political structures. Insurgency tactics include guerrilla warfare, terrorism, and political activism, such as propaganda and recruitment. The insurgency operations are designed to weaken government control and delegitimize it, debilitate, or reduce its presence in zones they control.[12] Four types of insurgencies generally challenge the state: political, military, traditional, and urban. The kind of insurgency will also dictate its strategies, territory, zones of operations, and leadership. Strategically, insurgencies function as organizations and must have clear objectives to achieve their political agenda.[13] Insurgent movements produce short-term results, but they are rarely successful. In the case of Colombia, the insurgencies embodied characteristics from all four types. FARC was a rural movement operating in the Colombian countryside, and M19—was an urban guerrilla operating mainly in the city, especially the administrative centers such as Bogota. The actions carried out by both groups had particularities based on the areas of operation. Urban insurgencies and rural insurgencies are both political. However, in the case of Colombia, FARC-EP had a military structure, whereas M19 was organized in operations units and recruited young students.

Areas containing FARC presence. [National Counterterrorism Center. https://www.dni.gov/nctc/ftos/farc_fto.html]

Multinational Operations and Counterinsurgency

Solving an internal conflict became a challenging task for the Colombian state. The dire situation and the pronged conflict made it quite impossible for the state to find the means to fight on different fronts without casualties. Counterinsurgency was one of the most challenging endeavors for the state, combatting at least three active armed insurgencies: FARC, ELN, and M19. During an armed conflict, the warring parties receive support from the populace supporting the cause and international interest groups, whether state or non-state actors. International support is essential for any movement. At the same time, any state with an internal conflict would seek support from various states or institutions—whether direct military assistance or diplomacy—to lessen and stop the conflict from spreading. The collaboration between states to combat insurgency is part of a particular military doctrine known as a multinational operation.

Multinational operations are conducted by forces of two or more nations, usually undertaken within the structure of a coalition or alliance. In a strategic context, cooperating nations form regional, geopolitical, and economic relationships to promote national interests, ensure mutual security against real and perceived threats, conduct foreign humanitarian assistance peace operations, and promote their ideals.[14]

When discussing the multinational cooperation between the U.S. and the Colombian state to combat the insurgency, it is paramount to consider the magnitude of the conflict, its complexity, the warring parties, and how it impacted national security. A series of factors prompted Plan Colombia. Among others was the threat FARC posed to the U.S. national security and the U.S. assets in the region. The U.S. assistance comprised counterinsurgency measures, reforming the penitentiary system, strengthening the military, the police force, and the air force, and introducing new technology in combatting both drug cartels and insurgents.[15] The most important tenet of the multinational operation between the U.S. and Colombia was redesigning COIN practices and measures through Plan Colombia.

The FARC

The 1960s marked a period where the communist movements consolidated throughout Latin America. The Cuban influence in the region was substantial, and the communists realized that traditional politics were insufficient to come to power.[16] Instead, a violent revolution was the only means to take over.[17] The Colombian conflict precedes the formal founding of the FARC on May 27, 1964.[18] Colombian political and social life has always suffered from polarization and violence stemming from inequalities and failure to apply democratic principles. After a brief military rule (1953–1958),[19] the two major parties (Liberals and Conservatives)—arranged a power-sharing agreement and formed a National Front Coalition. Under the agreement, the parties would rotate power every four years. This power-sharing agreement lasted for sixteen years, yet it did not settle the disagreement between the extremist factions within the parties.[20] The disagreements drove the left-wing factions into extremism and launched two insurgencies: The FARC–EP and the Ejercito de Liberacion Nacional (ELN) (1964). The guerrilla groupings emerged in stages. The first-generation clusters were the EPL (Ejército Popular de Liberación) and ELN.[21] In 1974, Movimiento Autodefensa Obrera and Partido Revolucionario de los Trabajadores emerged as urban guerrillas.[22] The insurgent groups operated in urban areas and the countryside. FARC and ELN shared the same totalitarian ideology, justifying totalitarianism and a full-blown revolution as beneficial and good. The leftwing insurgencies in Colombia competed with the state and the political elite for power. They also competed with the paramilitaries to hold on to the agricultural sector and vast territories.

During the years of violence, the PCC recruited many poor farmworkers from rural central Colombia in the provinces of Tolima and Cundinamarca.[23] The PCC operated as a mobilizing political structure to unite the peasants, directly threatening the Liberal and Conservative elite during the conflict. Hence, the PCC was an instrumental force in organizing the self-defense movement, and its leaders, Marulanda Velez and Jacobo Arenas were also PCC members and the founders of FARC–EP.[24]

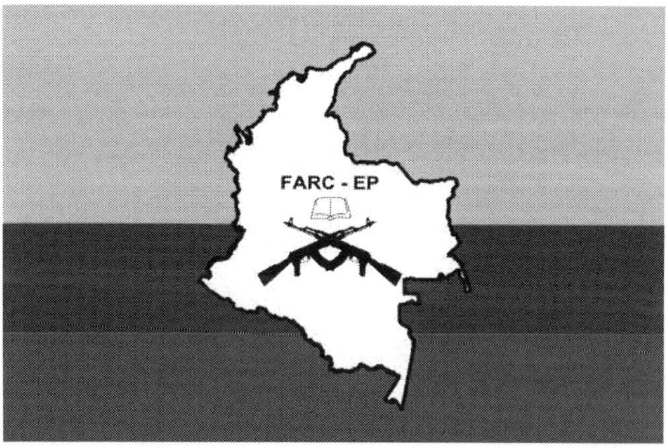

FARC-EP FLAG [National Counterterrorism Center. https://www.dni.gov/nctc/ftos/farc_fto.html]

Social Order Under Insurgency

As a belligerent party, the insurgency had many structures and levels and operated in rural and urban areas.[25] The rebels sought to debilitate and paralyze the state's presence in the disputed regions, establish rules where they were absent, dismiss the state's laws, and install a new order. This way, they would create some sense of normalcy. The insurgent factions (FARC and ELN) sought to form a parallel government in the controlled zones during the conflict.[26] Anna Arjona (2017), in her work *Rebelocracy: A Theory of Social Order in Civil War*, refers to it as life under rebelocracy. In such a setting, rebels seek order within the chaos; they control the infrastructure and local governance, including schools, hospitals, roads, and bridges. In other words, insurgency in Colombia threatened national security and sovereignty.[27]

Establishing a rebelocracy did not produce any political results for FARC M19 or ELN members. As an extreme faction of the PCC, they were out of Colombian mainstream politics. Even when a few former guerrilla fighters and leftwing politicians tried to join mainstream politics in the 1980s, they were assassinated. Such were Bernardo Jaramillo of Union Patriotica and Carlos Pizzarro Leongomez of the Democratic Alliance.[28] It was a failed revolution within a larger conflict that never translated into a transformative process. Although the most prolonged insurgency expanded in number and scope, it stalled in strategic planning and execution. Survival became a primary goal. With the fall of communism in the USSR (Union of Soviet Socialist Republics) and in Eastern Europe in the early 1990s, the ideology was no longer relevant to leftwing movements.

Losing the ideological axis transformed FARC from a grievance-based organization to a greed-based insurgency. In 2000, FARC sought to revise its ideology by founding a new political structure, Movimiento Bolivariano por la Nueva Colombia (The Bolivarian Movement for the New Colombia). It attempted to rekindle the organization's ideology, espousing Bolivar's universalism. The party dissolved in 2017 after FARC demobilized as part of the peace deal.

FARC engaged in terrorist acts, violent crimes, kidnapping, and assassinating members of the Colombian political elite.[29] During the conflict, 35,000 murders were committed only by the leftwing guerrillas, FARC, ELN, and M19.[30] FARC was, without question, a violent insurgency and operated in 242 areas, making up approximately 22 percent of Colombia's territory. It was a serious threat, a mobilized political and military force that competed with the state and the paramilitary. Recruiting tactics were pure propagandistic means to solicit the population affected the most by the agrarian law. Many of the members who joined were landless peasants and indigenous people.[31]

The presence of the state in the areas occupied by the movement was also an indicator of the scope of the operations. In the case of FARC–EP, the state's presence was questionable in the regions where FARC had the most popularity. FARC was the corollary of pent-up frustration with the agrarian reform and appealed to the lower strata—significantly displaced farmers. A CIA special report revealed that FARC was considered banditry and not a threat to national security and stability, at least in the early sixties.[32]

The guerrilla warfare in Colombia was diverse and complex. It had many clusters varying from militarized forces operating in the Colombian jungle to attacks by urban guerrillas in cities and institutions. M19,[33] a leftist guerrilla, was responsible for carrying out actions ranging from car bombing to kidnapping until the assault on the Justice Palace in 1985. Twelve Supreme Court Justices were killed, along with 48 Colombian soldiers. The rise of the guerrilla movement in Colombia resulted from two significant developments in the first half of the twentieth century.[34] First, it was conceived from political violence. The movement sought to transform itself into a revolutionary action. By creating a military force, the guerrillas aspired to exceed the preponderance of the state, combat it, and eventually supplant it. Secondly, the guerrillas rose as an auto-defense military cluster to protect the farmers and their territory. In either case, their ideal objective was to turn the war into a revolution with a set of ideological beliefs, take over the state, and eventually replace it with a new political regime.[35]

The FARC: Structure and Operations

"The guerrilla as a military organization is offensive, professionalized in its legal structure, and governed by elementary command rules," argued Jacobo Arenas, one of FARC's ideological leading figures.[36] Arenas was known for his aggressive tactics in challenging the state using guerrilla warfare, such as city sieges and direct combat. Apart from his ideological warfare view, Arenas also sought the revolution as the only means to overthrow the Colombian government and establish a regime run by insurgents.[37]

From a structural point of view, FARC was a hierarchical military organization. The ultimate power was vested in a six-member general secretariat under the leadership of Marulanda Velez. Structurally, FARC operated with blocs. Twenty-five higher commanders oversaw the blocs dispersed in various country areas. The blocks were then divided into fronts. The front was a ranking military unit, a guerrilla command operating in a specific geographic jurisdiction. A front included the combat troops, which made up its core.[38]

Apart from the combat troops, there were support and infrastructure elements. Usually, at the core of the front were the combat units, generally one or several companies. Every company consisted of 50-55 fighters. In addition, a company was made of two platoons or guerrillas, each having 25 soldiers. A further platoon subdivision was the squads with eight fighters. At the front level, various commissions such as finance—tax collection service, logistics, intelligence gathering, public order, and mass work commissions reassured that the insurgency was functioning and had sufficient funds for operating.[39] The communists in the rural communities sought to recruit and organize people contingent on guerrilla operations and sympathizers.[40]

The tactics were strictly guerrilla warfare, ambushing state military units and raiding farms to seize land from wealthy landowners. FARC sought to capture hostages and military equipment to secure food and supplies. Illegal mining also became a source of income.[41] Since 2002, the group has increased its capability to attack major cities. With the advent of the War on Terrorism, FARC could focus more on U.S. targets.[42]

Pedro Antonio Marin, also known as "Tirofijo" or "Manuel Marulanda", the top commander of the Revolutionary Armed Forces of Colombia (FARC). [AFP PHOTO/Rodrigo Arangua. https://www.denverpost.com/2008/05/24/top-rebel-reported-as-dead-in-colombia/]

In 1997, FARC entered the list of terrorist organizations. Until then, FARC was a threat to Colombia's internal stability. Per Rand.org, FARC increased its military operations during Plan Colombia: "...As discussed above, this expansion is closely correlated to perceived increased threats from the U.S. backed – Plan Colombia. Thus, we conclude that FARC has the resources and ability to manage its weapons supplies strategically and could threaten the United States should it choose to do so."[43]

At the beginning of the movement, FARC generated income mainly from taxes and serving as security units for the narco-traffickers, and then, when realizing that they had the resources and the infrastructure to trade cocaine, it seized the opportunity by directly engaging in the production and sales of cocaine. The drug trade now funded their agrarian revolution, yet FARC did not have a robust military strategy—except for occupying and operating in rural Colombia.[44] Later, the assistance from the U.S. gave Colombia's military a boost. As a result, FARC suffered losses and a decrease in numbers. Survival became a priority. To prohibit failure within, it established codes and rules.[45] "In 2000, the United States and Colombia initiated Plan Colombia, a $9 billion U.S. military aid program meant to help the Colombian government combat the drug trade, reassert authority, and increase its capacity throughout the country. ... However, some credit Plan Colombia with increasing the strength of the Colombian state and military, as well as initiating the FARC's decline."[46]

From Grievance-based to Greed-based Insurgency

The disintegration of FARC as an insurgency cannot be attributed only to the fading of communism globally. Instead, it was a slow internal process, a combination of internal processes and external factors, including the state's power, its monopoly over the use of force, various actors vested in the cause, and international support. Since its origin, FARC arose as an insurgency in response to the ongoing political conflict and the failure of a fair agrarian reform. Demagogically, FARC justified the means and their actions, conforming with Marxist theory and overlooking all democratic means to solve the crisis, yet sought to cash in drug money in the last decade, becoming notorious drug traffickers. Scholars, especially from a military background, distinguish two types of insurgencies in how FARC survived and operated—first, the grievance-based insurgency, and second, the greed-based insurgency. The grievance-based insurgencies became popular during the Cold War. When communism was established in various parts of the world, the disenchanted and marginalized population was suddenly left out or, in many cases, persecuted for religious, ideological, or political convictions. The bitterness was enough reason to fight or overthrow systems and governments.[47] In a polarized world such as during the Cold War era, grievance-based insurgencies were supported by both the United States and the Soviet Union.[48] While the U.S. foreign policy consisted of stopping communism from exporting in other regions of Latin America, the Soviet Union backed left revolutions or insurgencies such as in Cuba and Chile. Even in countries where it did not bring about a communist revolution, the Soviet Union supported the communist parties created before or after the 1940s.

To fight a substantial war, insurgencies must have widespread popular support, especially in domestic conflicts. It is the center of gravity for an insurgency to thrive and survive. Thus, despite support from key international actors, the insurgents must find resources and self-sponsor the war.[49] The key to its success is the number of fighters it enlists and a steady flow of income and resources. Grievance and ideology are the most efficient recruitment motivators, and propaganda is the tool for success. Even the most rogue insurgency has its rhetoric: "Whatever the cause might be, it is always the right one, and the government is the oppressor."[50]

In the case of greed-based insurgency, there are underlying assumptions that illicit activities, looting, extortion, drug trafficking, and other profiting trades or businesses make a substantial income to finance the war. Greed-based insurgencies do not seek to overthrow a government but limit their scope to a) having access to illicit activities and b) having fewer clashes with the government. In this case, greed-based insurgencies do not have a clear ideology.

In the case of FARC, its nature is more complex, and FARC embodied characteristics from both defections. Hence, as a grievance-based insurgency, FARC had a clear ideology based on the Marxist concept of primitive accumulation and class warfare and demanded that a revolution solve the agrarian problem. As for the greed-based insurgency—FARC used the means to accumulate substantial wealth from the drug trade to sponsor the agrarian revolution and maintain thousands of guerrilla soldiers enlisted in the movement. The intent of FARC as a greed-based insurgency was not at the same level as the drug cartels because FARC operated on ideology and propaganda. While drug cartels were after the state to obstruct the extradition law, FARC was a military structure operating as a shadow government in the regions where the state's presence was not significant. It could run entire areas, control their infrastructure, schools, and hospitals, and run an illicit economy.[51]

FARC's means were scathing to the people of Colombia who became victims during the conflict and challenging for the Colombian state since it operated parallel for decades. FARC revealed a deep social wound. It exposed a divided society ravaged by violence and bad politics. It also showed how fragile the state becomes while sovereignty and jurisdiction over the national territory are in peril.

The Necessity of Plan Colombia

Throughout the 20th Century, Colombia has witnessed many insurgencies, some still active. One of the reasons for this unusual form of political discontent was low tolerance in identity politics and political impasse. The precursor of the Colombian armed conflict was the Bogotazo, a violent outbreak in Colombia in 1948 between the liberals and conservatives over the assassination of Jorge Eliecer Gaitan, a liberal politician. At that time, the political turmoil in Colombia was a corollary of the lack of consensus and the desire to solve the agrarian reform through democratic means. Over the years, the conflict developed many complexities and spread across the country. In the late 1990s, the state had exhausted its human resources and military capacity. The state was involved in multiple fronts with non-state actors such as guerrillas and drug cartels that cooperated amongst each other or against each other. The guerrilla movement was backed by autocratic regimes or ideological insurgencies such as the Sandinistas, the Cuban and Venezuelan governments, and even the Russian government.

The urgency to assist Colombia in combatting the insurgency was paramount for the United States. The conflict had utterly changed the fabric of Colombian society. People were born in violence and grew old with the violence. A RAND publication, "Victory Has a Thousand Fathers," details the reasons and factors why counterinsurgency measures fail. As mentioned above, fighting insurgent factions is one of the most challenging endeavors for a state, especially when the insurgency is highly ideological and has support from the populace. The populace affected by the conflict and in support of the insurgency is the key to determining the success of COIN.

Apart from the insurgency having control over parts of the national territory, the zones of guerrilla operations converted into coca labs for the drug industry. The conflict had many layers. The guerrillas fought the state and the paramilitary at the same time. The state, on the other hand, faced two violence-producing entities involved in drug trafficking. Drug cartels were openly against the Colombian state's justice system and funded campaigns to influence the politics and the extradition law.

Rural Colombia and the red zones became battlefields where the state and FARC guerrillas often exchanged fire. FARC mined their controlled land to keep the state and the paramilitary out. The population in these zones lived in dire situations. Various interest groups and political adversaries, including the Colombian political elite, were divided on many issues. The extradition law—was one of the critical issues since it alleviated some of the pressure in domestic politics. It impacted the financing and the income of the warring groups in the conflict: the insurgents, the state, and the drug cartels.

During the Clinton Administration, the approach to assisting Colombia in solving the crisis was primarily diplomatic. The United States' foreign policy did not consider the conflict a threat to U.S. sovereignty. After drug trafficking increased in the late 1990s, Colombia's internal strife became a stability threat in the region. Designing and implementing Plan Colombia became a necessity. FARC and ELN have always been opponents of the state and endangered Colombia's internal stability and sovereignty. The funds generated from drug trafficking funded FARC and its operations. Before entering the drug trade, FARC generated income mainly from taxes and served as security units for drug traffickers. Realizing that they had the resources and the infrastructure to trade cocaine, they seized the opportunity by directly engaging in the production and sale of cocaine.[52]

The U.S. assistance to Colombia's military began in the 1940s. The United States trained Colombian army officers in the U.S. military schools, boosting their capacity, especially with the rise of the insurgency. The long-time cooperation comprised many areas, and the U.S. support poured onto Colombia's national security to invigorate the fight against the narcotics and the left-wing rebels. The military agreements between the U.S. and Colombia started right after WWII, and it soared in the early 1950s with the deployment of a Colombian combat battalion to the Korean War. During the Korean War, Colombian soldiers fought along with the U.S. units.[53]

ELN fighters in Catatumbo [TV San Jorge: "Cifras Del Conflicto Que Estremecen El Catatumbo." YouTube, July 29, 2019. https://www.youtube.com/watch?v=4AtwbQgEvww.]

Plan Colombia was a United States foreign, military, and diplomatic initiative to combat Colombian drug cartels and left-wing insurgencies. It was initiated by Andres Pastrana, the Colombian President (1998–2002), and the U.S. President Bill Clinton, and it was passed in Congress in 2000. By the early 2000s, the U.S. and the Colombian military had increased their cooperation in combating narco-trafficking. The U.S. Special Forces continued to boost the Colombian military by providing technology and devices used in air traffic. In addition, the U.S. Air Force radar sites assisted in aircraft tracking and shared information with the Colombian military. The U.S. Special Forces helped Colombian police and the DEA intercept illegal activity, especially the drug laboratories in the jungle.

The joint operations between the U.S. and Colombia continued well from 2001 to 2007. With the new technology, the Colombian military, guided by the U.S. and Special Operation Forces, could precisely locate FARC's activity. Yet, capturing the rebels was challenging at times. While the detection of FARC training camps was precise, they dispersed as soon as they detected activity.

To deter and decrease insurgent activity, the U.S. provided helicopters and tested the smart bomb for the first time. Yet they were only used with U.S. approval. The Pentagon held the codes to the smart bombs as a secret to avoid being misused to eliminate political opponents or for any other faulty intentions. The U.S.'s role in shifting Colombian politics cannot be diminished or denied. For decades, military assistance created more prepared military leaders with clear strategies and objectives. These strategies became paramount in decreasing FARC's capacities and operations, including the dismantling of drug cartel networks.

Findings and Analysis

This study sought to analyze how ideological insurgencies function and operate as political and military entities in opposition to a democratic or semi-democratic system.

There is a positive correlation between ideology and insurgency. It moves upward when ideology is strong and downward when ideology is weak. The upward trend comprises solidifying the movement, more substantial support, the proliferation of guerrilla fighters, increased resources and tax collection, expanded control over the territories, and conducting well-planned military operations. The insurgency justifies violence to solve grievances. In the case of FARC, and as I discussed above, the political faction (FARC–EP) was an illicit organization—yet *de facto* still part of the PCC, including its leaders. FARC aimed to challenge the state at a military level. In contrast, the Communist Party of Colombia challenged the system in mainstream politics. Even though they operated as two entities—FARC and PCC had the same political objective and ideology- they used different mechanisms to address grievances. FARC–EP, M19, and ELN used violence and justified it for the greater good.

The role that leftwing propaganda played in the Colombian conflict was influential. It enabled the mobilization and recruitment of new members. With the help of the PCC, FARC was able to increase its numbers. PCC would provide FARC with the ideological basis, political leadership, strategies, and assistance in its occupied zones. From the 1960s until the 1990s, during the Cold War era, the Eastern Bloc, including the USSR, had built close ties and alliances with the communist parties in Latin America, such as Cuba, Brazil, Colombia, and Venezuela. The zeitgeist of that period was to create a multipolar world order to counterbalance the West's military and economic power. The turmoil of the 1990s in Eastern Europe and the Soviet Union caused a shift in the balance of power in favor of the West. Hence, the collapse proved that the leftwing ideology that kept the systems together was flawed. The decline was a blow to leftwing insurgencies.

Financing the movement remained one of the biggest challenges that FARC–EP faced. Rising as a revolutionary army, the FARC used the ideology to gather support for the cause. Founding a shadow government in mostly rural Colombia enabled the rebels to create a system of collecting taxes and governing the occupied zones. Establishing rebelocracy meant securing some normalcy in the FARC-controlled areas while the cash flow continued.

In conclusion, the conflict in Colombia could not be alleviated without the help and assistance it received from the U.S. Hence, international support was paramount to deter left-wing insurgencies and drug cartels. U.S. military assistance enabled Colombia to strengthen its institutions and boost its confidence. Colombia remains a U.S. partner in the War on Terrorism, yet the political situation remains fragile and problematic. ELN is still active, and there are reported paramilitary activities in Santa Marta.[54] Illegal activities and various armed groups are still present in the region. The Colombian government is currently showing more leniency toward drug trafficking, making it again the most profiting industry in the area—exceeding earnings from oil exports.[55]

Endnotes

1 The Revolutionary Armed Forces of Colombia, or FARC, was one of the most prolonged insurgencies of the twentieth century. FARC was founded in 1964 as an extremist faction of the Colombian Communist Party.

2 Rand.org. (n.d.). Arms trafficking and Colombia - Rand corporation. https://www.rand.org/content/dam/rand/pubs/monograph_reports/2005/MR1468.pdf

3 U.S. Army Special Operations Command (USASOC) History Department. "The 1960s." The 1960s: A Decade of Revolution. Accessed September 12, 2023. https://arsof-history.org/articles/v4n4_1960s_page_1.html.

4 Ibid.

5 Lee Taylor is a Georgetown University's School of Foreign Service and Allen Walter student. "Case Analysis: The FARC in Colombia." Case Analysis: The FARC in Colombia | *Small Wars Journal.*

6 Colombian Communist Party was founded in 1930 as a Comintern substructure – for Colombia. It is a legal structure under the leadership of Jaime Caycedo. The central ideology is Marxism, Leninism, & Bolivarianism.

7 Maria Del Pilar Lopes Uribe, and Fabio Sánchez Torres. *On the Agrarian Origins of Civil Conflict in Colombia*. Accessed January 17, 2024.

8 Charles Tilly, The *Politics of Collective Violence*. Cambridge University Press, 2003, 26.

9 Byman, D. L., Chalk, P., Hoffman, B., Rosenau, W., & Brannan, D. (n.d.). Trends for outside support for insurgent movements. https://www.rand.org/dam/rand/www/external/congress/terrorism/phase2/insurgent.pdf

10 Nazih Richani, *Systems of Violence: The Political Economy of War and Peace in Colombia*. Albany: Suny Press, 2013.

11 Mark Youngman, University of Birmingham e-Data Repository. Accessed September 13, 2023.

12 Guide to the Analysis of Insurgency. Federation of American Scientists. Accessed September 14, 2023. https://irp.fas.org/cia/product/insurgency.pdf.

13 Ibid. CIA pg. 4.

14 Multinational Operations – Joint Chiefs of Staff. Accessed September 19, 2023. https://www.jcs.mil/Portals/36/Documents/Doctrine/pubs/jp3_16.pdf.

15 Colombia: Background and U.S. Relations - CRS Reports. Accessed September 19, 2023. https://crsreports.congress.gov/product/pdf/R/R43813.

16 The Cuban Revolution gave hope to communist guerrillas in Colombia to seek other means to seize power. While the PCC supported the guerrillas, it did not directly engage in violence to maintain its legal status and participate in elections. Banditry and Insurgency in Columbia.

17 Ibid, p. 6.

18 *MMP: Revolutionary Armed Forces of Colombia (FARC)*. CISAC. (n.d.). https://cisac.fsi.stanford.edu/mappingmilitants/profiles/revolutionary-armed-forces-colombia-farc

19 Richard L. Maullin, Defense Technical Information. Center Soldiers, Guerrillas and Politics in Colombia. 1974.

20 Ibid.

21 Ejercito de Liberacion Nacional – a political structure founded around the same time as FARC in 1966 – it still active and operates in various zones of Colombia, such as Cauca, Antioquia, and Venezuela.

22 Partido de los Trabajadores (Workers Party), a political structure seeking to overthrow the state and establish a Marxist-Leninist regime. Erich Saumech Cadavid.

23 Partido Communista Colombiano (Colombian Communist Party). Pedro Antonio Marin was one of the movement's leaders. Later, he changed his name to Manuel Marulanda Velez and became the supreme commander of FARC. Manuel Marulanda Velez was born in 1930 and died in 2008.

24 Gary Leech, *The FARC: The longest insurgency*. London: Zed Books Ltd, 2011.

25 Banditry and Insurgency in Columbia. The World Factbook. Accessed September 20, 2023. https://www.cia.gov/readingroom/docs/DOC_0000598514.pdf.

26 NCNC. "The Methods and the Tactics of Global Terrorism." Counterterrorism Guide - National Counterterrorism Center. Accessed January 19, 2024. https://www.dni.gov/nctc/index.html.

27 Ibid.

28 Bernardo Jaramillo was a former FARC guerrilla, presidential candidate for Union Patriotica, and Carlos Pizarro Leongomez was a former M19 rebel presidential candidate for the Democratic Alliance. Both candidates were assassinated before the election in 1990 by the paramilitary.

29 Colombia Reports. "FARC Claim Responsibility for Killing One of Colombia's Most Legendary Politicians." Colombia News | Colombia Reports, October 5, 2020. https://colombiareports.com/amp/farc-claim-responsibility-for-killing-one-of-colombias-most-legendary-politicians/.

30 In Sight Crime. "FARC." In Sight Crime, October 20, 2022. https://insightcrime.org

31 Banditry and Insurgency in Columbia. The World Factbook. Accessed September 20, 2023. https://www.cia.gov/readingroom/docs/DOC_0000598514.pdf.

32 Ibid, p. 7.

33 M19 – a military guerrilla group that operated in urban areas of Colombia – especially in Bogota, was responsible for the assault on the Palace Justice in 1985.

34 Erich Saumeth Cadavid. Historia de la guerrilla en Colombia. Accessed September 20, 2023. https://www.didacticamultimedia.com/registro/estudios/10/documentos/guerrilla_colombiana.pdf.

35 Ibid.

36 "*La guerrilla es, según su definición, "una organización militar de carácter ofensivo profesionalizada en su estructura jurídica y regida por normas elementales de comando.*" Jacobo Arenas (original text).

37 2014, Autor de Kaos. "Colombia: A 20 Años de La Muerte Del Camarada Jacobo Arenas, Su Biografía." Kaos en la red, August 11, 2011. https://archivo.kaosenlared.net/colombia-a-20-a-os-de-la-muerte-del-camarada-jacobo-arenas-su-biograf-a/.

38 Christine Sixta Rineheart, Volatile *social movements and the origins of terrorism: The Radicalization of Change*. Lanham, MD: Lexington Books, 2013, 80.

39 Angel Rabasa and Peter Chalk, *Colombian labyrinth the Synergy of drugs and insurgency and its implications for regional stability*. Santa Monica, CA: RAND, 2001.

40 Reuters. "Colombia Army Raids Illegal Mines Funding FARC Rebels." Reuters, May 11, 2015. https://www.reuters.com/article/colombia-mining-illegal-idINKBN0NW27820150511.

41 Ibid.

42 Rand.org – In a special report about the situation in Colombia after Plan Colombia was implemented – FARC guerrillas increased members and recruiters to counterbalance the state's power. FARC could attack American assets in the region.

43 Ibid.

44 Kim Cragin and Bruce Hoffman, *Arms Trafficking and Colombia.* Rand Corporation. Accessed September 24, 2023. https://www.rand.org/content/dam/rand/pubs/monograph_reports/2005/MR1468.pdf.

45 Ibid.

46 Ibid.

47 Self, K. A. (n.d.). Counterterrorism Policy in Columbia.

48 Ibid.

49 Ibid.

50 Paul Collier and Anke Hoeffler, *Greed and Grievance in Civil War.*

51 Ana Arjona, *Rebelocracy: Social Oorder in the Colombian Civil War.* Cambridge, U.K: Cambridge University Press, 2017.

52 Michelle Hughes and Michael Miklaucic, "Colombia and the FARC: From Military Victory to Ambivalent Political Reintegration?" Essay. In *Impunity: Countering Illicit Power in War and Transition.* Washington, D.C.: Center for Complex Operations, 2016.

53 Dana Priest, "Covert Action in Colombia." *The Washington Post.* Accessed September 23, 2023. https://www.washingtonpost.com/sf/investigative/2013/12/21/covert-action-in-colombia/.

54 Colombia Reports. "The Wars for Colombia's Cocaine Containers: Part 3: Santa Marta." Colombia News | Colombia Reports, April 25, 2022. https://colombiareports.com/the-wars-for-colombias-cocaine-containers-part-2-santa-marta/.

55 Oscar Medina, "Cocaine Is Expected to Overtake Oil as Colombia's Main Export This Year." Bloomberg.com, September 14, 2023.

Bibliography

2014. Autor de Kaos. "Colombia: A 20 Años de La Muerte Del Camarada Jacobo Arenas, Su Biografía." Kaos en la red, August 11, 2011.

Arjona, Ana. *Rebelocracy: social order in the Colombian civil war.* Cambridge, U.K: Cambridge University Press, 2017.

Banditry and Insurgency in Colombia. The World Factbook. Accessed September 20, 2023. https://www.cia.gov/readingroom/docs/DOC_0000598514.pdf.

Byman, Daniel L., Peter Chalk, Bruce Hoffman, William Roseneau, and David

Brannan. Trends for outside support for insurgent movements. Accessed September 12, 2023. https://www.rand.org/dam/rand/www/external/congress/terrorism/phase2/insurgent.pdf.

Cadavid, Erich Saumech. Historia de la guerrilla en Colombia. Accessed September 20, 2023. https://www.didacticamultimedia.com/registro/estudios/10/documentos/guerrilla_colombiana.pdf.

Collier, Paul, and Anke Hoeffler. Greed and Grievance in the Civil War. Accessed September 24, 2023. https://documents1.worldbank.org/curated/en/192611468769173749/pdf/multi-page.pdf.

Colombia: Background and U.S. Relations – CRS Reports. Accessed September 19, 2023. https://crsreports.congress.gov/product/pdf/R/R43813.

Colombia Reports. "FARC Claim Responsibility for Killing One of Colombia's Most Legendary Politicians." *Colombia News* | Colombia Reports, October 5, 2020.

Colombia Reports. "The Wars for Colombia's Cocaine Containers: Part 3: Santa Marta." *Colombia News* | Colombia Reports, April 25, 2022.

Cragin, Kim, and Bruce Hoffman. Arms trafficking and Colombia – Rand Corporation. Accessed September 24, 2023. https://www.rand.org/content/dam/rand/pubs/monograph_reports/2005/MR1468.pdf.

"Declassified Documents Constitute Key Evidence in Judgement against Colombian Paramilitary." National Security Archive, December 24, 1997.

DeMare, Brian James. *Land Wars: The story of China's Agrarian Revolution*. Stanford, CA: Stanford University Press, 2019.

Guide to the Analysis of Insurgency. Federation of American Scientists. Accessed September 14, 2023. https://irp.fas.org/cia/product/insurgency.pdf.

Hughes, Michelle, and Michael Miklaucic. "Colombia and the FARC: From Military Victory to Ambivalent Political Reintegration?" Essay. In *Impunity: Countering Illicit Power in War and Transition*. Washington, D.C.: Center for Complex Operations, 2016.

Human Rights Watch: Informe Anual 1998 (Eventos de 1997) — Colombia. Human Rights Watch | informe anual 1998 (eventos de 1997) Colombia. Accessed September 24, 2023. https://www.hrw.org/legacy/spanish/inf_anual/1998/colombia.html.

In Sight Crime. "FARC." InSight Crime, October 20, 2022.

Leech, Gary. *The FARC: The Most Prolonged Insurgency*. London: Zed Books Ltd, 2011.

Maullin, Richard L. Defense Technical Information. Center Soldiers, Guerrillas and Politics in Colombia.

"MMP: Revolutionary Armed Forces of Colombia (FARC)." CISAC.

Medina, Oscar. "Cocaine Is Expected to Overtake Oil as Colombia's Main Export This Year." Bloomberg.com, September 14, 2023.

Multinational Operations – Joint Chiefs of Staff. Accessed September 19, 2023. https://www.jcs.mil/Portals/36/Documents/Doctrine/pubs/jp3_16.pdf.

Priest, Dana. "Covert Action in Colombia." *The Washington Post*. Accessed September 23, 2023. https://www.washingtonpost.com/sf/investigative/2013/12/21/covert-action-in-colombia/.

Rabasa, Angel, and Peter Chalk. *Colombian Labyrinth: The Synergy of Drugs and Insurgency and its Implications for Regional Stability*. Santa Monica, CA: RAND, 2001.

Richani, Nazih. *Systems of violence: The Political Economy of War and Peace in Colombia*. Albany: Suny Press, 2013.

Rinehart, Christine Sixta. *Volatile Social Movements and the Origins of Terrorism: The Radicalization of Change*. Lanham, MD: Lexington Books, 2013.

Self, Kevin A. Counterterrorism Policy in Columbia. Accessed September 23, 2023.

Staff, Reuters. "Colombia Army Raids Illegal Mines Funding FARC Rebels." *Reuters*, May 11, 2015.

"A Supreme Court Judge Whose Predecessor Was Assassinated Was... - UPI Archives." UPI, October 30, 1986.

"The Friends of 'El Viejo': Declassified Records Detail Suspected Paramilitary, Narco Ties of Former Colombian President Uribe." National Security Archive, September 23, 1991. https://nsarchive.gwu.edu/briefing-book/colombia/2020-08-31/friends-el-viejo-declassified-records-detail-suspected-paramilitary-narco-ties-former-colombian.

Tilly, Charles. *The Politics of Collective Violence*. Cambridge University Press, 2003.

"Timeline: U.S.-Colombia Relations." Council on Foreign Relations. Accessed September 24, 2023. https://www.cfr.org/timeline/us-colombia-relations.

U.S. Army Special Operations Command (USASOC) History Department. "The 1960s." The 1960s: A Decade of Revolution.

Youngman, Mark. Ubira E-theses – University of Birmingham e Data Repository. Accessed September 13, 2023. https://etheses.bham.ac.uk/id/eprint/9372/7/Youngman2019PhD.pdf.

Crusader Vows

Paul M. Aitchison
Florida College

Abstract

Crusader vows during and after the Crusading Period changed due to the politization of the religious and solemn vow. The concept of crusading expanded beyond its original intention as the Levantine Crusades ended and the Crusader States failed. Originally used as a sign highlighting to others the free choice in going on a pilgrimage to Jerusalem in the form of a cloth cross, potentially proclaimed the willingness to fight for God. The military orders pledged vows and wore crosses. Later, the vow meant one was willing to fight against Muslims, heretical Christians, and pagans in Eastern Europe. The origins of the vow continue today in the form of the military oath taken by commissioned officers of many militaries.

Keywords: Middle East, Medieval, Military, Religious, Western Europe, Eastern Europe, Baltics

Juramentos de los cruzados

Resumen

Los juramentos de los cruzados durante y después del Período de las Cruzadas cambiaron debido a la politización del voto religioso y solemne. El concepto de cruzada se expandió más allá de su intención original cuando terminaron las Cruzadas Levantinas y fracasaron los Estados Cruzados. Originalmente utilizado como un signo que resaltaba a los demás la libre elección de ir en peregrinación a Jerusalén en forma de cruz de tela, proclamaba potencialmente la voluntad de luchar por Dios. Las órdenes militares hacían juramentos y llevaban cruces. Más tarde, el juramento significó que uno estaba dispuesto a luchar contra los musulmanes, los cristianos heréticos y los paganos en Europa del Este. Los orígenes del juramento continúan hoy en la forma del juramento militar que hacen los oficiales comisionados de muchos ejércitos.

Palabras clave: Medio Oriente, Medieval, Militar, Religioso, Europa Occidental, Europa del Este, Países Bálticos

doi: 10.18278/sshj.12.1.4

十字军誓言

摘要

由于庄严的宗教誓言的政治化，十字军时期和之后的十字军誓言发生了变化。随着黎凡特十字军东征的结束和十字军国家的失败，十字军东征的概念超出了其初衷。此概念最初被用作一个标志，以布十字架的形式向其他人强调前往耶路撒冷朝圣的自由选择，其潜在表明“为上帝而战”的意愿。军事团体对此宣誓并佩戴十字架。后来，这个誓言意味着愿意与穆斯林、异端基督徒以及东欧的异教徒作战。此誓言的起源至今仍以“许多军队长官宣誓誓词”的形式延续。

关键词：中东，中世纪，军事，宗教，西欧，东欧，波罗的海

This paper examines Crusader vows during the major Crusades into the Levant and Holy Land (1090s AD–1291 AD). During this period, the impression and perception of the crusading movement expanded beyond its original intent. The battle of Manzikert in 1071 in eastern Asia Minor saw the Byzantium army defeated and the last of the Diogenes family of emperors captured. This disaster spurred the call to the west for aid. The Council of Piacenza occurred in the spring of 1095. Among the topics of discussion between ecclesial and laypeople was a Byzantine request for knights to help him restore the lost territories of Anatolia. The call for aid was not the first time the Eastern Roman Empire had asked for assistance from the west. This distress signal came when the Roman Catholic Church and the Eastern Orthodox Church had better relations after the Schism of 1054. The Byzantine Emperor Alexius I Comnenus (1081–1118) was restored from being excommunicated after Pope Urban II (r. 1088–1099) came to the papal throne. The historical ground was set, with examples from the recent past where Popes looked to the early Church fathers and Doctors of the Church, such as Saint Augustine of Hippo (354–430 AD) for the justification of arming individuals to fight for the cross. As the Levantine Crusades ended and the Crusader States failed, the Church and the idea of taking a crusading vow also underwent a change. The lineage of this vow continues in secular ceremonies such as the military officers oath. The notion of crusading and what it would take for an individual brought to go on an armed pilgrimage was not something that developed out of thin air.[1]

Previous popes in the early eleventh century, such as Pope Gregory VII (1073–1085), looked for ways to push the profession of arms from a realm of the laity to the realm of the spiritual.

The Peace of God and Truce of God, beginning in the tenth century, was at first regionally focused, with bishops in their respective diocese alongside local secular leaders in the Frankish kingdom attempting to keep nobility of the area from attacking paupers on the roads and not to steal. The procession of relics, alongside the masses at the local parish level added pressure to local magnates to assist in cleaning up the roadways and encourage trade. As masses began to be more formalized in the eleventh century, the notion that daily masses could do "good" also became more prevalent. As a result, the Church in Europe was seen as the authority for spiritual and sometimes secular punishment for those who fought against that "goodness." The types of punishments under the authority to be used by local bishops included public penance, interdict and formal oaths and vows, which always had a religious formula. The local and regional synods occurring in the tenth and eleventh century addressed much of this to provide not just a legal proceeding against unnecessary violence but also with creating solutions to the problems alongside the secular leaders. An example of this would be in the Synod of Poitiers in 1000 and 1014, which recommend a peace militia be created for the purpose of enforcing the Peace and True of God. Taking a more militant approach by clerical leaders to secular problems began to occur throughout the eleventh century with the crusades or the armed pilgrimage being a outcome, only realized decades later. The Peace of God and the Truce of God was not able to stop the violence, and in the decades following the localized movement towards a more general peace the Papacy took steps towards formalizing violence away from Europe and from those who lacked the ability to protect themselves.[2]

The focal point of what became the First Crusade (1095–1099) was not on being led by the church but led by nobles and knights of the west who would take a vow from the clergy which laid out the bounds of what was allowed during the pilgrimage. Urban II's speech is lost to the time of history; several write-ups occurred in the subsequent decades and centuries that became the foundation of crusading and taking a vow before God. In that sermon, it is assessed by the chroniclers who wrote after the event that Urban encouraged the clergy to meet their responsibilities to their local parishes and diocese, to correct themselves and to agree to enforce the Truce of God.[3]

> "The Lord Pope said also, 'Brothers, you must suffer for the name of Christ many things, wretchedness, poverty, nakedness, persecution, need, sickness, hunger, thirst, and other such troubles, for the Lord said to his disciples, 'You must suffer many things for my name.' And 'Be note ashamed to speak before men, for I will give you what you shall say and afterwards 'Great will be your reward.' And when these things had begun to be rumored abroad throughout all the duchies and counties of the Frankish lands, the Franks,

> hearing them, straightaway began to sew the cross on the right shoulder of their garments, saying that they would all with one accord follow in the footsteps of Christ, by whom they had been redeemed from the power of hell."[4]

The first challenge the Papacy needed to overcome was sanctioning the use of force. Gregory VII laid the foundation for using force by merging crusading with armed pilgrimage and holy war. The combination of papal approval, the laity's vow to God and the church, and the holy war for Christ created the conditions that Urban II used to preach the First Crusade throughout France and Germany in the late eleventh century. A unifying holy war with an armed pilgrimage to support the church's cause on Earth was a powerful force. The vow, taken by the laity in front of the clergy, reinforced the religious and sacrificial aspect of the violence committed on behalf of religion. The council of Clermont, instituted to manage a number of problems for the Church in Western Europe at the end of the eleventh century, would be remembered as a council not of peace but one of war, with the institution of war being directed against the infidel outside of Christendom in the name of the Church.[5]

When Pope Urban II in 1095 called for an armed pilgrimage to retake Jerusalem from the Seljuk Turks, he asked for those promising to affirm with a solemn vow before a priest and God. An outward sign of this affirmation was the attachment of a cloth cross to their outer garment right after the ceremony. The cross was the sign of the willingness of the person to risk his life in battle for God. The cloth cross, for centuries, gave physical protection to crusaders against persecution, search, and secure and legal foreclosure. Urban II was looking for an opportunity, as a medieval feudal lord, to extend his control over a more substantial portion of Europe and thus focused his words during the preaching of the First Crusade on freeing the eastern churches from the clutches of not only the Muslim Caliphates and sultanates but also the Eastern Orthodox Church. In the spontaneous atmosphere, Urban II and his clerical administration also unveiled their greatest gift, one used sparingly in the past, the granting of indulgences. The clergy's goal for the First Crusade and the knightly class's goals would soon clash.[6]

Saint Bernard, at this time the abbot of Clairvaux (1090–1153), wrote to the Master of the Knights Templar, Hugh de Payens, around 1128 AD, extolling the virtues of pleading one's life to the militant cause of fighting for God. In the aftermath of the success of the First Crusade, the collapse of the County of Edessa against the Muslim counteracts. At the core of the original success of the First Crusade was the preaching, guided by Pope Urban II. For the Second Crusade, Pope Eugene III (r. 1145–1153) commissioned the greatest orator of the twelfth century, Bernard to not only preach before other clergy to train them on their roles and responsibilities of preaching the Crusade to the laity, he also took time to lay down the rules for the military order of the Knights Templar, fusing the idea of a militant monk fighting against the forces aligned against God.

> "Go forth confidently then, you knights, and repel the foes of the cross of Christ with a stalwart heart. Know that neither death nor life can separate you from the love of God which is in Jesus Christ, and in every peril repeat, "Whether we live or whether we die, we are the Lord's." What a glory to return in victory from such a battle! How blessed to die there as a martyr! Rejoice, brave athlete, if you live and conquer in the Lord; but glory and exult even more if you die and join your Lord. Life indeed is a fruitful thing and victory is glorious, but a holy death is more important than either. If they are blessed who die in the Lord, how much more are they who die for the Lord!"[7]

The granting of indulgences provided the means by which the church could recruit a large number of soldiers for the task, and the ceremony by which to formally send those soldiers outside of Western Christendom needed to be sufficiently sacred. The vow, a religious obligation to the church as a whole, was at once religious and also militant. Even the symbol, the cloth cross, a sign of the ultimate sacrifice by the king of kings and leader of the world, was a military symbol for the army gathering in Europe to fight and to continue to do so in the face of peril. While the record of Urban II preaching concerning the First Crusade is incomplete, it is asserted through other sources, including letters of abbeys and monasteries in France and the Holy Roman Empire that Urban wanted those who could actually fight to go east, and wanted all to have an oath affirmed by local clergy to fight for the liberty of Christians in the east.[8]

If Urban had used the goal of freeing Jerusalem from Muslim control, then naturally total focus for the armed pilgrimage would have been to rid the city of God. However, it does not seem to be the case. Instead, Urban emphasized at the beginning of the preaching the importance of freeing fellow Christians in Asia Minor from the yoke of oppression in the east. Only afterward, as Ademar, the bishop of Le Puy and the papal legate of the crusades, continued to shape the narrative of what the armed pilgrimage would morph into, freeing Jerusalem became the shining goal. In addition, his immediate predecessor, Gregory VII, began to formulate the idea of a *militia santci Petri,* who served as the vicar of Christ on Earth, or the papal throne first and foremost. The idea would continue to evolve into the military orders, the Knights Templar and Hospitallers, and the Teutonic Order in the early 12th century.[9]

The recruitment for the First Crusade was very much an *ad hoc* affair. While the Papacy preached the first Crusade, they were not the ones who were the military or logistical leaders of the armies; that fell to the nobles and military commanders. The papal legate for the Crusade could suggest, extoll, and encourage the troops toward the goal but could not make liege men fight. The methods used by Urban II to preach the First Crusade in providing indulgences and protection from prosecution. Fulcher of Chartres writes that at Clermont, Pope Urban stated,

> "Let those accustomed to waging private wars wastefully even against Believers, go forth against the Infidel in a battle worthy of being undertaken now and to be finished in victory ... let those, who was recently hired for a few pieces of silver, with their eternal reward."[10]

The original formulation in the mind of Gregory VII and Urban II might have been to aid Constantinople because of the pleas from the Byzantine emperor not only to the papal court but also to the nobles in the Frankish and German lands. The laity, whipped to a frenzy by the preaching of wandering itinerate priests centuries before the Franciscans' formation, may have played a role in how the Pope laid out why and who should go on the armed pilgrimage to the east. It is likely, especially in the aftermath of the Papal Schism of 1054, that Urban II saw the armed expedition to the east as being under the overall command of the Greek Emperor to bring about a healing between the two branches of Christianity and would be fighting in Asia Minor with a distant and long after attaining the goal of liberating the whole of the Levant and Jerusalem for Christianity. From the records of Clermont, it is apparent that the movement's overall goal was the liberation of Christians and the capture of Jerusalem.[11]

The naming of the First Crusade did not occur until centuries had passed, and several of what was initially termed as penitential armed pilgrimages to the east had already been completed. The crusading ideal, coming straight from the throne of Saint Peter, was a sanctioned, holy war against those who opposed the church by individuals sanctioned by that same church to do violence in their name for the regeneration of faith on Earth and to bring about the final judgment of God on the whole Earth. The evangelizing efforts of the wandering preachers were thought to bring about a general reawakening in the spiritual aspects of the commoners and nobles alike in Europe.[12]

In the decades up to the First Crusade, the church continued to evolve its stance on violence and the sanctioning of violence on behalf of the church. One bishop, Adalbero of Laon, expressed it in the following terms: The House of God, which was believed to be one, was in reality three because, in it, three categories of men performed three distinct functions, the first prayed, the second fought and the third worked. The church began to see the need to combine the functions after the success of the First Crusade, nobles who participated in the crusades also began to see the need to create a military order that would combine the aspects of praying and fighting to provide a solid military force for the new cru-

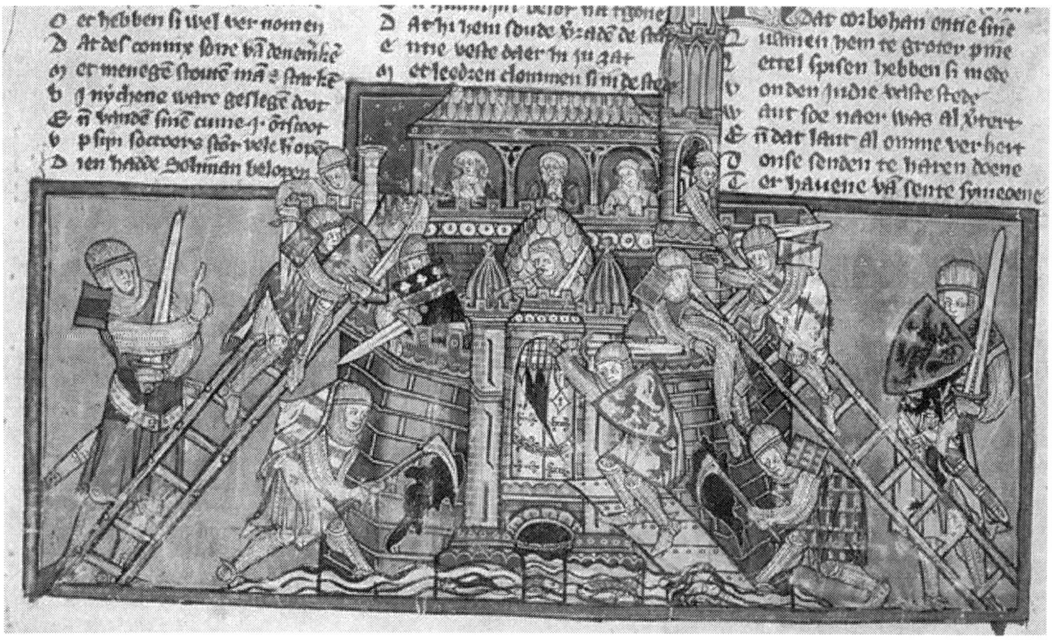

A 14th century CE illustration depicting the siege of Antioch of 1097-1098 CE during the First Crusade. Some crusaders forsook their vows and departed the Levant. Those who stayed conquered the city. [National Library of the Netherlands, The Hague.]

King Louis IX devoted much of his adult life to crusading, taking the cross twice for the 7th and 8th Crusade. He also served temporarily as the King of Jerusalem. [Illustration of Louis IX from manuscript BL Royal 16 G VI, f. 404v, British Library]

sading states in the east. These so-called holy knights, or the military orders included the Poor Fellow-Soldiers of Christ and of the Temple of Solomon or the Knights Templar (1119), and Order of Knights of the Hospital of Saint John of Jerusalem, or the Knights Hospitaller (1099) would also provide safe passage to the lucrative pilgrimage trade of individuals and goods moving back and forth across the Mediterranean and Europe as the crusades evolved.[13]

The vow of a crusader was different than other vows taken in religion because ever since the First Crusade, it was done in conjunction with an indulgence provided by the local bishop through the Pope, and while it was done at the individual level, it was completed through the efforts of many people working together. It was a vow before God but also a public vow before other Christians, in one aspect to keep each other accountable and in another to be pushed forward as a collective to an ultimate earthly, secular goal, the capture of territory, a city, or the liberation of oppressed Christians in a region. While general vows before the clergy might be for penitence or to press someone publicly to complete an arduous task or labor, the crusading vows focused the goal of those taking the vow in the direction of violence against fellow men.[14]

The church always wanted to have a larger role in the planning and execution of the crusades, especially after the success of the First Crusade, and attempted to do so through the *crucesignati* or those wearing the sign of the cross—to make a solemn vow committing themselves to the endeavor proposed by Popes as they called for renewed offensives against their enemies. As the vow itself was comprised of a personal statement of intent to carry out a certain series of actions on earth in an attempt to gain rewards in the life after. The acts crusaders hoped to complete martial acts rarely seen in Europe, to march to the Middle East, and take over large areas of defended cities and citadels. The fact that although the vow was a reciprocal pact made between an individual and God, the Latin Church acted as God's guarantor. Within that structure, the pope had a large stake in creating the terms of the crusader vow in such a manner that would benefit the Papacy as a whole. Those who took the vow had to fulfill the requirements to receive the spiritual benefits, and this combined the penance of pilgrimage with acts of violence to be assured of the reward. Nevertheless, this cannot and should not overshadow the fact that the vow was still an agreement between the individual and God—one which was oftentimes personalized to reflect one's own understanding of this activity. Crusaders and those we who organized and promoted the crusade in the first century in which crusading took place possessed different understandings of what it meant to be signed with the cross.[15]

In the enthusiasm of the First Crusade (1095–1096), Second Crusade (1147–1149) and the Third Crusade (1189–1192), some individuals desperately wanted to fulfill their vow but were physically unable to do so for whatever reason. The church, at the time, created a substitute for the penitential exercise of going on the Crusade, and one could provide a monetary donation instead. Not only did this provide a sacrifice in the minds of the church, but it would also provide funds for someone who might be more physically able to pursue the crusading vow to travel and fight. This continued to morph and evolve as other crusades had different goals and motivations, and the personal and societal understanding and opinion of crusading changed. The evolution of the crusading vow brought with it problems of abuse and coopting of the religious and solemn nature to be subordinate to the martial and secular nature. St. Bernard, in laying out the rules of the Templar, makes an example of the life of a crusader under a vow.[16]

> "In the first place, discipline is in no way lacking and obedience is never despised. As Scripture testifies, the undisciplined son shall perish and rebellion is as the sin of witchcraft, to refuse obedience is like the crime of idolatry. Therefore, they come and go at the bidding of their superior. They wear what he gives them, and do not presume to wear or to eat anything from another source. Thus, they shun every excess in clothing and food and content themselves with what is necessary. They live as brothers in joyful and sober company, without wives or children. So that their evangelical perfection will lack nothing, they dwell united in one family with no personal property whatever, careful to keep the unity of the Spirit in the bond of peace. You may say that the whole multitude has but one heart and one soul to the point that nobody follows his own will, but rather seeks to follow the commander."[17]

The abuses of the crusading vow began as soon as the First Crusade was preached. Individuals, who might be in dire legal or financial situations, would immediately take the cross and gain the privilege and protection. They would then look for ways to keep that protection while not actually having to fulfill the vow, thus creating the abuse of liberty. As more and more crusades were preached through the twelfth and thirteen centuries, the Papacy began to develop a system of sending out preachers who were trained in what to say and who to look for to start to form the army of a crusade. The author of the *Gesta Francorum* states that,

> "For even the Pope set out across the Alps as soon as he could, with his arch-bishops, bishops, abbots, and priests and he began to deliver eloquent sermons and to preach, saying, "If any man wants to save his soul, let him have no hesitation in taking the way of the Lord in humility, and if he lacks money, the divine mercy will give him enough."[18]

Strong-willed popes, such as Innocent III (r. 1189–1216), also harangued nobles and kings alike in west-

A romantic view of Peter the Hermit Preaching the First Crusade. Note the crowds of classes of society pleading and pledging to go on the armed pilgrimage – [Cassell's History of England, Vol. I - anonymous author and artist]

ern Europe to coordinate the fulfillment of vows in a way that would increase his temporal power while displaying the might of the Papacy in the aftermath of the setbacks in the east and the fall of Jerusalem and most of the Crusader States. Innocent also brought a youthful vigor to the throne of Saint Peter, being 37 years old when he ascended to the head of the Christian world in the west. Innocent's youthful exuberance for crusading, both near and far across Western Europe and the Levant was only outstripped by his ability to cajole, recruit and finance the crusading movement. In the encyclical, *Quia maior,* calling for the Fifth Crusade, Innocent III called for the reclamation of the Holy Land because it belonged to the Christian faith. He utilized apocalyptic and eschatological language in describing the "beast was approaching" and that almost 666 years elapsed since Muhammad's Hijrah. This was after he already called the Fourth Crusade, the Albigensian Crusade, and the Iberian Crusade.[19]

The military orders, the Knights Templar, and the Knights Hospitaller pledged solemn vows and wore cloth crosses that let all know who they fought for in Palestine. In the subsequent generations of crusaders, the vow meant one was willing to fight against Muslims in Spain, Christian heretics in southern France, and pagans in eastern Europe and the Levant. The vow, alongside the indulgence offered by the clergy, was a powerful motivation to send men, women, and whole families to the farther reaches of the known world at that time to fight for Christ.

Alongside the military orders was the secular knightly order of society, those who fought. By the eleventh century, the *milites* were attracting the attention of the most prominent institution, the church. The church's primary focus was on the spiritual and morale aspect of the warrior society, to focus the soldier's wrath against a target that would benefit all of Christendom and shift the focus away from the disruption to the daily life and rhythm of western Europe. The need for a defender for the church was apparent before the First Crusade. At the local diocese level, bishops and abbots attempted to reduce the violence through the Peace of God and other such efforts to curb the enthusiasm of the combatants. The church understood that they could not stop wars, and even some wars were to the benefit of the Papacy through the protection and expansion of land holdings but, more importantly, in the safety of the sanctity of the clergy. The combination of the clerical with the militant was a potent one indeed.[20]

KNIGHT IN ARMOUR.
MS. Roy. 2 A. xxii.
Late Thirteenth Century.

Kneeling crusader knight. The military orders used the cross extensively on their knightly garb. A religious vow could not be forsaken by a secular ruler. [13th-century illustration from Westminster Psalter. British Library]

The churches focus on the hierarchy of things on the Earth directly related to how they viewed the cosmic scale. As a result, they saw and believed the role of the soldiers as one of defense and defense of the church. The gradual process after the First Crusade in developing and establishing the military orders was melding the sacred with the secular, the holy, and the profane or wrathful.[21]

Thus, in addition to being soldiers, those who took the cross were also pilgrims. Crusaders also solemnly vowed to travel to Jerusalem and offer their prayers at the Holy Sepulchre. Observers and participants alike understood the First Crusade first and foremost as a pilgrimage. One sees this no more clearly than by looking at the form of the vow taken by crusaders in

Urban II at Clermont. Urban preached what became the First Crusade mainly to other prelates of the church and nobles, who in turn preached in their own realms. Note the extensive use of crosses on both the secular and spiritual garb. [unknown artist, mid-14th century]

1095 AD and thereafter. The crusade vow mimicked the vow made by devotional pilgrims in the period leading up to the First Crusade with two important additions. First, as one pronounced the vow a cloth cross was attached to the votary's garment, and secondly, "benediction was bestowed on swords along with [pilgrim's] staffs" in recognition that unlike other pilgrims the crusaders were allowed, in fact encouraged, to carry weapons while undertaking their penance. If one was to allow for any consensus in regard to how participants saw the crusade it would have to be based upon this: to one extent or another every crusader agreed that the crusade was a peculiar form of pilgrimage.[22]

The vow, given with all solemnity and before an ordained priest of the Roman Catholic Church, was, first and foremost, a religious ceremony. Moreover, as a religious ceremony, it is still seen today in the evolution of our legal system in swearing in a witness, and the United States oath of officer for the branches of the Armed Forces. Religion brings people worldwide together in discourse, faith, and, unfortunately, conflict. The High Middle Ages are seen as a simple yet dark time by many who are not historians of that era. However, western Europe's culture, religion, and society were advancing in a rapid pace, with the developing ideas of nations, regions, and communities beginning to bring about an awareness of the wider world. The connection of a vow, combined with the benefit of an indulgence to undertake the quest to the Holy Land, was in the fulfillment in the eyes of the medieval church and secular society, a sign of God's grace. As Jerusalem fell back into Muslim hands, the crusades and the crusading vow began to take on a new importance, as Gregory VIII noted that the crusades were a test to see if people were prepared to do penance and risk their lives in service of something higher than themselves. Also, Innocent III spoke about the Crusade as the service that God requires to show obedience to him and in failing to accomplish the goals of the various crusades, the crusaders themselves were being tested and either coming out stronger or weaker. The importance of having a standard for the world, built on a foundation of something greater.

The military oath has a long and rich history dating back centuries. Its importance and growth have evolved over time, and it is still a vital part of the military today. The oath serves as a binding contract between soldiers and their country, and it has deep religious roots. The concept of oaths can be traced back to ancient civilizations, where it was believed that the gods would punish those who broke their promises. In the Middle Ages, knights took oaths of loyalty to their lords and to the church. The first recorded military oath was taken by the Roman army in 312 BC. The Roman soldiers swore an oath of allegiance to their commander and to the state. This oath was taken very seriously, and those who broke it were punished severely. The Christian notion of vows also played a significant role in the development of the military oath. In the Bible, Jesus teaches that one's word should be binding, and that a person should not make oaths lightly. Christians believed that oaths should be made before God, and that breaking them was a sin. This belief carried over into the military, where soldiers would take oaths in the name of God. Over time, the military oath changed and evolved.[23]

During the American Revolution, soldiers took an oath of loyalty to the United States, swearing to defend the country against all enemies, foreign and domestic. The oath was updated again during the Civil War, when soldiers were required to swear an oath of allegiance to the Constitution. Primary source material from the Civil War era provides insight into the importance of the military oath. In a letter to his wife, Union soldier John A. Mather wrote, "I take the oath of allegiance to the United States of America with my whole heart and soul. I am ready to fight and die for my country."[24] This sentiment was echoed by many soldiers on both sides of the conflict. During World War II, the military oath was updated once again. The modern military oath, which is still used today, includes a pledge to obey the orders of the President and those appointed over the soldier. It also includes a promise to uphold the Constitution and to defend the country against all enemies, foreign and domestic. Religious aspects of the military oath are still present today. Many soldiers choose to take the oath with their hand on a Bible, and the oath itself includes the phrase "so help me God." However, soldiers of all faiths and those who do not have a particular religious affiliation may take the oath in a way that is meaningful to them. In conclusion, the military oath has a long and rich history, with deep religious roots. It has evolved over time to reflect the changing nature of warfare and the role of the military in society. The importance of the military oath cannot be overstated, as it serves as a binding contract between soldiers and their country. Primary source material from throughout history provides valuable insight into the meaning and significance of the military oath.[25]

Crusading vows proved a motivational force in sending Christians from thousands of miles across Europe together to gain Jerusalem for Christianity and fulfill what they saw as their duty to God in furthering a religious cause on the Earth and bringing about a judgment against those seen as enemies of God. The vow called upon something greater than the military force on the ground to assist in the defense or offense. It provided a spiritual defense to both the common soldier departing his family for the final time, the noble with his large retinue and retainers, and the clergy, ministering to the spiritual needs of those on the crusade. The physical defense also protected the property of those away from their families and homesteads. As it continued to evolve, crusaders taking vows could go crusading for a season, or a few years or for a campaign, and then return home. Swearing a vow before a host of people was also a way to bring a group of people together as a team, and that is through that history the military oath used today evolved. It serves as a commitment to uphold the values and principles of the armed forces, serves as a reminder of the high expectations and responsibilities placed on service members today and reinforces the idea of unity and teamwork. The evolution of crusading vow shaped the conflicts between Christianity and Islam to this day.

Endnotes

1 D.C. Murno, "The Speech of Pope Urban II at Clermont, 1095," *American Historical Review,* xi (1906): 231-242.

2 Frederich Kempf, *The Church in the Age of Feudalism,* (New York: Crossroads, 1982), 346.

3 Monumenta Germaniae Historica. *Electronic Monumenta Germaniae Historica: Emgh.* München Turnhout: Monumenta Germaniae Historica: Brepols, 2002), 462; *The Historia Ierosolimitana of Baldric of Bourgueil.* (Woodbridge: Boydell Press, 2014), 124-125; Georg Strack," The Sermon of Urban II in Clermont and the Tradition of Papal Oratory." *Medieval Sermon Studies,* vol. 56 (2012): 32-33.

4 St. Bernard of Clairvaux, *In Praise of the New Knighthood*, trans. By M. Conrad Greenia (Trappist, Kentucky: Cistercian Publications, 1977), 1-2.

5 St. Bernard of Clairvaux, *In Praise of the New Knighthood*, trans. By M. Conrad Greenia (Trappist, Kentucky: Cistercian Publications, 1977), 23; Paul A. Chevedden, "Crusade Creationism "versus" Pope Urban II's Conceptualization of the Crusades," *The Historian*, 75 (2013), 25-29; Peter Hecht, "Crusading as Philosophical Construct: Thoughts and Actions of Pope Urban II, St. Bernard, and Peter the Venerable," *Comparative Civilizations Review*: Vol. 84: No. 84, Article 6, (2021): 37-41; Giovanni Dominicus Mansi, "Quia qui christianum occidit, sine dubio christi sanguinem fundit," in *Sacorum Concilliorum Nova,* vol. 19, col. 827 (Lepizig, F.M. Geidel, 1903), 536-542; Jean Flori, "Knightly Society." In *The New Cambridge Medieval History,* ed. David Luscombe and Jonathan Riley-Smith, (Cambridge: Cambridge University Press, 2004), 165.

6 Cowdrey, H.E.J, "Pope Urban II's preaching of the First Crusade," *History*, Vol. 55, No. 184 (1970): 178-180; John Patrick Doland, *Handbook on Church History: The Church in the Age of Feudalism*, (The University of Michigan: Herder and Herder, 1969), 448-449.

7 Bernard Bernard. *Treatises III: On Grace and Free Choice in Praise of the New Knighthood*. Kalamazoo Mich: Cistercian Publications, 1997, 2-3.

8 John Patrick Doland, *Handbook on Church History: The Church in the Age of Feudalism*, (The University of Michigan: Herder and Herder, 1969), 448-449.

9 H.E.J. Cowdrey, "Pope Urban II's preaching of the First Crusade," *History*, Vol. 55, No. 184 (1970): 178-180.

10 Fulcher of Chartres, *Chronicle of the First Crusade,* trans. M.E. McGinty (Philadelphia: University of Philadelphia Press): 1941, 15-17; Jonathan Riley-Smith, "The Crusades, 1095–1198." Chapter. In *The New Cambridge Medieval History*, edited by David Luscombe and Jonathan Riley-Smith, 4: 534–563. *The New Cambridge Medieval History*. Cambridge: Cambridge University Press, 2004, 536.

11 Jonatha Riley-Smith, "The Crusades, 1095–1198." Chapter. In *The New Cambridge Medieval History,* edited by David Luscombe and Jonathan Riley-Smith, 4: 534-563. *The New Cambridge Medieval History*. Cambridge: Cambridge University Press, 2004, 537-539.

12 Tyerman, Christopher, *God's War: A New History of the Crusades,* (Harvard: Harvard University Pressm 2009), 72; H.E.J Cowdrey, "Pope Urban II's Preaching of the First Crusade," *History*, 55 (1970): 188; Paul E. Chevedden, "The Islamic View and the Christian View of the Crusades: A New Synthesis." *History,* vol. 93, no. 2 (31) (April 2008): 181-187.

13 Adalberro of Laon, "Triplex ergo Dei domus est, quae creditor Una;/Nunc orant, alii pugnant, aliique laborant," *Carmen ad Rotbertum regem*, (Italy, 1760), 297-8; Jean Flo-

ri, "Knightly Society." In *The New Cambridge Medieval History,* ed. David Luscombe and Jonathan Riley-Smith, (Cambridge: Cambridge University Press, 2004), 165.

14 Christoph T. Maier, "The Friars and the Redemption of Crusade Vows." Chapter. In *Preaching the Crusades: Mendicant Friars and the Cross in the Thirteenth Century,* 135-160. Cambridge Studies in *Medieval Life and Thought: Fourth Series.* Cambridge: Cambridge University Press, 1994, 134-135.

15 T.S. Asbridge, *The First Crusade: A New History* (New York: 2004), 43-45; Jennifer Ann Price, "Cruce Signatus: The Form and Substance of the Crusading Vow." (PhD diss, University of Washington, 2005), 272-273.

16 Christoph T. Maier, "The Friars and the Redemption of Crusade Vows." Chapter. In Preaching the Crusades: Mendicant Friars and the Cross in the Thirteenth Century, 135-160. Cambridge Studies in Medieval Life and Thought: Fourth Series. Cambridge: Cambridge University Press, 1994,135-136.

17 Bernard Bernard. *Treatises III: On Grace and Free Choice in Praise of the New Knighthood.* Kalamazoo Mich: Cistercian Publications, 1997, 4-6.

18 *The Deeds of the Franks and the Other Pilgrims to Jerusalem.* (London: T. Nelson, 1962), 48.

19 Christoph T. Maier, "The Friars and the Redemption of Crusade Vows." Chapter. In Preaching the Crusades: Mendicant Friars and the Cross in the Thirteenth Century, 135-160. Cambridge Studies in Medieval Life and Thought: Fourth Series. Cambridge: Cambridge University Press, 1994, 139-40; Hehl, Ernst-Dieter. "War, Peace and the Christian Order." Chapter. In *The New Cambridge Medieval History,* edited by David Luscombe and Jonathan Riley-Smith, 4: 185-228. *The New Cambridge Medieval History.* Cambridge: Cambridge University Press, 2004, 188; Vatican City, Archivio Segreto Vaticano, *Registra Vaticana* 8, fos. 140v-41v; edited in *Patrologiae Cursus Completus,* Series Latina, ed. J.P. Migne (221 vols., Paris, 1844-65) ccxvi, vols. 817-22; Tyerman, Christopher, *How to Plan a Crusade,* (New York: Pegasus Books), 219; Tyerman, Christopher, *God's War: A New History of the Crusades,* (Harvard: Harvard University Press, 2009), 612; T.W. Smith, "How to craft a crusade call: Pope Innocent III and *Quia maior* (1213). *History Research,* vol. 92. Iss. 255 (2019); 6.

20 Jean Flori, "Knightly Society." In *The New Cambridge Medieval History,* ed. David Luscombe and Jonathan Riley-Smith, (Cambridge: Cambridge University Press, 2004), 149, 165; Giovanni Dominicus Mansi, "Quia qui christianum occidit, sine dubio christi sanguinem fundit," in *Sacorum Concilliorum Nova,* vol. 19, col. 827 (Lepizig, F.M. Geidel, 1903), 536-542.

21 Hehl Ernst-Dieter, "War and peace and the Christian order," in *The New Cambridge Medieval History IV,* (Cambridge: Cambridge University Press, 2004), 209-210.

22 Ekkehard of Aura, *Chronicon,* ed. Joseph Stevens (Vaduz: Kraus Reprint, 1964), 214;

Geoffrey of Villehardouin, *La conquete de Constantinople,* ed. E. Fara!, Les Classiques de l'Histoire de France au Mayen Age 18-19, 2 vols. (Paris, 1938-39), 1: 44; A. Gottlob, *Kreuzablass und Almosenablass: eine studie uber die Fruhzeit des Ablasswesens* (Amsterdam, 1965), 308; Jennifer Ann Price, "Cruce Signatus: The Form and Substance of the Crusading Vow." (PhD diss, University of Washington, 2005), 272-275.

23 Rosa, John W., *Preface to the U.S. Air Force Academy Officer Development System,* Jan 2004, 1; Reese, Thomas, "An Officer's Oath," *Military Law Review,* 25 (July 1964): 3; *The Armed Forces Officer's Guide,* Department of Defense, 2006, 4.

24 Mather, John Austin, *Journals, 1832-1880,* Andover-Harvard Theological Library, 2012, 144.

25 Hehl Ernst-Dieter, "War and peace and the Christian order," in *The New Cambridge Medieval History IV,* (Cambridge: Cambridge University Press, 2004); 209-210; Gerhoh of Reichersberg, *De aedificio Dei*, chap. 43; *PL* 194, cols. 1300--4. (Regensburg: Schnell and Steiner, 2020), 103; Rosa, John W., *Preface to the U.S. Air Force Academy Officer Development System,* Jan 2004, 1; Reese, Thomas., "An Officer's Oath," *Military Law Review,* 25 (July 1964): 3; *The Armed Forces Officer's Guide,* Department of Defense, 2006, 4.

Bibliography

Ann Price, Jennifer., "Cruce Signatus: The Form and Substance of the Crusading Vow." PhD diss, University of Washington, 2005.

Asbridge, T.S., *The First Crusade: A New History* Oxford: Oxford University Press, 2004.

Chartres, Fulcher of., *Chronicle of the First Crusade,* trans. M.E. McGinty Philadelphia: University of Philadelphia Press 1941.

Chevedden, Paul A., 'Crusade Creationism "versus" Pope Urban II's Conceptualization of the Crusades', *The Historian*, 75 (2013): 1-46.

Chevedden, Paul E. "The Islamic View and the Christian View of the Crusades: A New Synthesis." *History,* vol. 93, no. 2 (31) (April 2008): 181-200.

Cowdrey, H.E.J, "Pope Urban II's preaching of the First Crusade," *History,* Vol. 55, No. 184 (1970): 177-188.

Doland, John Patrick., Handbook *on Church History: The Church in the Age of Feudalism*, The University of Michigan: Herder and Herder, 1969.

Ekkehard of Aura, *Chronicon,* ed. Joseph Stevens Vaduz: Kraus Reprint, 1964.

Flori, Jean., "Knightly Society." In *The New Cambridge Medieval History,* ed. David Luscombe and Jonathan Riley-Smith, Cambridge: Cambridge University Press, 2004.

Gottlob, A., *Kreuzablass und Almosenablass: eine studie uber die Fruhzeit des Ablasswesens* Amsterdam, 1965.

Hehl Ernst-Dieter, "War and peace and the Christian order", in *The New Cambridge Medieval History IV,* Cambridge: Cambridge University Press, 2004.

Hecht, Peter., "Crusading as Philosophical Construct: Thoughts and Actions of Pope Urban II, St. Bernard, and Peter the Venerable," *Comparative Civilizations Review*: Vol. 84: No. 84, Article 6, (2021): 37-41.

Kempf, Frederich., *The Church in the Age of Feudalism,* New York: Crossroads, 1982.

Laon, Christoph T., 'Triplex ergo Dei domus est, quae creditor Una;/Nunc orant, alii pugnant, aliique laborant', *Carmen ad Rotbertum regem*, Italy, 1760.

Maier, Christoph T., "The Friars and the Redemption of Crusade Vows." Chapter. In *Preaching the Crusades: Mendicant Friars and the Cross in the Thirteenth Century*, 135160. Cambridge Studies in *Medieval Life and Thought: Fourth Series.* Cambridge: Cambridge University Press, 1994.

Mather, John Austin, *Journals, 1832-1880,* Andover-Harvard Theological Library, 2012.

Mansi, Giovanni Dominicus., "Quia qui christianum occidit, sine dubio christi sanguinem fundit," in *Sacorum Concilliorum Nova,* vol. 19, col. 827, Lepizig, F.M. Geidel, 1903.

Murno, D.C., "The Speech of Pope Urban II at Clermont, 1095," *American Historical Review,* xi (1906): 231-242.

Monumenta Germaniae Historica. *Electronic Monumenta Germaniae Historica: Emgh.* München Turnhout: Monumenta Germaniae Historica: Brepols, 2002.

Reichersberg, Gerhoh of. *De aedificio Dei*, chap. 43; *PL* 194, cols. 1300-4. Regensburg: Schnell and Steiner, 2020.

Reese, Thomas., "An Officer's Oath." *Military Law Review,* 25 (July 1964): 3-17.

Riley-Smith, Jonathan., "The Crusades, 1095–1198." Chapter. In *The New Cambridge Medieval History*, edited by David Luscombe and Jonathan Riley-Smith, 4: 534-563. *The New Cambridge Medieval History*. Cambridge: Cambridge University Press, 2004.

Rosa, John W., *Preface to the U.S. Air Force Academy Officer Development System,* Jan 2004.

Smith. T.W. "How to craft a crusade call: Pope Innocent III and *Quia maior* (1213)." *Historical Research*, vol. 92, iss. 255 (2019): 1-23.

Strack, Georg., "The Sermon of Urban II in Clermont and the Tradition of Papal Oratory." *Medieval Sermon Studies,* vol. 56 (2012): 30-45.

St. Bernard of Clairvaux, *In Praise of the New Knighthood,* trans. By M. Conrad Greenia Trappist, Kentucky: Cistercian Publications, 1977.

The Armed Forces Officers Guide, Department of Defense, 2006.

The Historia Ierosolimitana of Baldric of Bourgueil. Woodbridge: Boydell Press, 2014.

The Deeds of the Franks and the Other Pilgrims to Jerusalem. London: T. Nelson, 1962.

Tyerman, Christopher, *God's War: A New History of the Crusade,* Havard: Harvard University Press, 2009.

Tyerman, Christopher, *How to Plan a Crusade: Religious War in the High Middle Ages.* New York: Pegasus books, 2017.

Villehardouin, Geoffrey of., *La conquete de Constantinople,* ed. *E. Fara!, Les Classiques de l'Histoire de France au Mayen Age* 18-19, 2 vols. Paris, 1938.

Cancer: On the Development of Chemotherapy

Demianna King
American Public University

Abstract

Understanding the contributions of Goodman, Gilman, and Farber in the development of chemotherapeutic agents is crucial to understanding the evolution of cancer treatment. Until very recently, cancer often meant a long and painful death sentence with no real answers. By investigating the progression of cancer treatment from antiquity into the modern age, society will ideally have a better idea of where to go next. Further research may unearth gaps in treatment, as well as additional questions as to how the medical community can continue to target various cancers more closely, extending quality and length of life.

Keywords: cancer, chemotherapy, Goodman, Gilman, Farber, chemical warfare, World War I, mustard gas, World War II, nitrogen mustards

Cáncer: sobre el desarrollo de la quimioterapia

Resumen

Comprender las contribuciones de Goodman, Gilman y Farber en el desarrollo de agentes quimioterapéuticos es crucial para comprender la evolución del tratamiento del cáncer. Hasta hace muy poco, el cáncer a menudo significaba una sentencia de muerte larga y dolorosa sin respuestas reales. Al investigar la progresión del tratamiento del cáncer desde la antigüedad hasta la era moderna, idealmente la sociedad tendrá una mejor idea de hacia dónde ir a continuación. Investigaciones futuras pueden descubrir brechas en el tratamiento, así como preguntas adicionales sobre cómo la comunidad médica puede continuar abordando varios tipos de cáncer más de cerca, extendiendo la calidad y la duración de la vida.

Palabras clave: cáncer, quimioterapia, Goodman, Gilman, Farber, guerra química, Primera Guerra Mundial, gas mostaza, Segunda Guerra Mundial, mostazas nitrogenadas

doi: 10.18278/sshj.12.1.5

癌症：论化疗的发展

摘要

理解古德曼、吉尔曼和法伯在化疗药物开发中的贡献，对于理解癌症治疗的演变至关重要。直到最近几年，癌症往往意味着漫长而痛苦的死刑，没有真正的答案。通过研究从古代到现代的癌症治疗进展，社会在理想情况下将更好地了解下一步该怎么做。未来研究可能会发现治疗方面的差距，以及关于"医学界如何继续更紧密地研究不同癌症，以延长生命质量和寿命"等其他问题。

关键词：癌症，化疗，古德曼，吉尔曼，法伯，化学战，第一次世界大战，芥子气，第二次世界大战，氮芥类

The development of chemotherapy, pioneered by Doctors Goodman, Gilman, and Farber, marked a pivotal moment in medical history with far-reaching impacts. Prior to their groundbreaking work, cancer was a disease with limited treatment options and bleak prognoses. However, Goodman, Gilman, and Farber's innovative approach to cancer treatment challenged conventional wisdom and transformed the way physicians approached the disease. By harnessing the power of chemical agents to target and destroy cancer cells, they provided a viable alternative to surgery and radiation therapy. This breakthrough not only extended the lives of cancer patients but also improved their quality of life by mitigating the debilitating symptoms associated with the disease. Moreover, the development of chemotherapy opened doors to further research and innovation in oncology, spurring advancements in drug discovery, delivery mechanisms, and personalized treatment regimens.

Cancer treatment in antiquity had a long-standing role as a disease with few treatment options and was difficult to heal from. Prior to the early 20th century, cancer was almost always a death sentence, and though death was not always immediate, it was certain. The search for a safe and effective treatment for cancerous tumors has been on the minds of physicians since the beginning of recorded history. One such document is the Edwin Smith Papyrus which presumably dates to around 3000-2500 BC.[1] Detailing 48 medical case studies including tumors of the breast, Smith's translation of this ancient papyrus offers valuable insight into the earliest known cancer treatments and how providers would have evaluated, treated, and described these ailments.[2]

Tactile examination of the described tumors was conducted to deter-

Alfred Gilman: Doctor Alfred Gilman working in a laboratory. [Yale School of Medicine. For faculty - past, present and future. Photograph. Jan-Feb 2009. https://medicine.yale.edu/news/medicineatyale/article/for-faculty--past-present-and-future/

mine size, shape, texture, temperature (to the touch and to evaluate for fever), and color to describe the state of the tumor and breast.[3] Depending on the findings of the physician, surgical techniques like cauterization, herbal salves, and bandages might then be applied to the tumor or open wound on the breast to facilitate healing. Per *The Edwin Smith Surgical Papyrus,* the course of treatment would vary depending on whether it was a "hot or cold tumor," and many of their herbal remedies, poultices, and salves contained ingredients like mason's mortar, acacia and sycamore leaves, ox dung, and copper salts.[4] Holistic treatments such as these set the standard for centuries to come with little change until the Renaissance, and many are even still in use today.

In the 16th to 18th centuries autopsies, advancements in surgical techniques and anesthesia allowed for rapid improvement in surgical treatment and "classic cancer operations such as the radical mastectomy were developed."[5] As surgical techniques improved, there was still relatively little treatment success due to post-operative infections and excessive bleeding during surgery. Anesthesia would revolutionize surgery by allowing surgeons to take more time, creating a more precise surgeon, and aseptic surgical fields and post-operative bandaging would greatly increase the safety of a surgical treatment option. Despite the growing safety of surgery, it would still be considered a last resort for some time with physicians continuing to search for ways to decrease surgical mortality rates. As Johnson and Lombard discuss in their 1941 paper, consideration needed to be taken to determine whether a surgical approach was truly in the best interest of the patient.[6] Their concern was that comorbidities like diabetes and obesity drastically increased the risks of anesthesia and their subsequent recommendation was not to perform any surgical procedures on those who would not recover well from the use of anesthesia.

The discovery of X-rays by Wilhelm Conrad Rontgen in 1895 opened a new chapter in medical science, catalyzing advancements in diagnostic imaging and therapeutic interventions.[7] Emil Grubbe, inspired by Rontgen's breakthrough, pioneered the application of X-rays in cancer treatment, marking a significant milestone in oncology. Grubbe's innovative use of X-rays to target and destroy cancerous cells in the late 19th and 20th centuries laid the foundation for modern radiation therapy.[8] Despite the primitive nature of early X-ray equipment, Grubbe's pioneering efforts yielded promising results and offered hope to cancer patients previously deemed untreatable. Unfortunately, early radiation therapy was imprecise and dangerous because researchers had not yet realized that radiation could cause cancer as often as it could cure it.[9] A tragic discovery considering many of these scientists and researchers would later develop leukemia due to their repeated use and testing of radiotherapy machines on themselves.[10]

With an ever-evolving need for more treatment options, research into the earliest chemotherapeutic drugs began in the 1910s.[11] With the onset of World War I, concerns of chemical warfare fueled further research into the potential therapeutic benefits of mustard gas. Initially used by Germany during World War I, sulfur mustard gas had devastating effects on the soldiers who were exposed, with 14,278 deaths reported in the first three weeks of its use.[12] In response to the newfound information, Yale University was given a contract from the U.S. Office of Scientific Research and Development to conduct further research into the potential therapeutic benefits of mustard gas.[13]

In 1941 two leading pharmacologists, Doctors Alfred Gilman and Louis Goodman began what would become some of their most celebrated work. Given the instability of sulfurous mustard gases for laboratory study, nitrogen mustards took their place. With further study it was concluded that nitrogen mustards "readily gained access to the circulation after topical application to the skin or following inhalation of vapor."[14] The observed cytotoxic activity of the nitrogen mustards on lymphoid tissue then encouraged Goodman and Gilman to begin clinical work on lymphosarcoma in mice. Though tumor growth would inevitably resume after treatment ceased in the mice, the results were promising enough to justify a clinical trial on malignant lymphoma, and so in the fall of 1942, Goodman and Gilman did just that.[15] This first trial was conducted at Yale University, and the first person to undergo treatment experienced a phenomenal and rapid remission despite being near death when the trial began. Much like the mice, these results would not last, however, the results were still encouraging enough for other researchers from the University

Louis Goodman: Doctor Louis Goodman posing for a headshot. [Yale School of Medicine. For faculty - past, present and future. Photograph. Jan-Feb 2009. https://medicine.yale.edu/news/medicineatyale/article/for-faculty--past-present-and-future/]

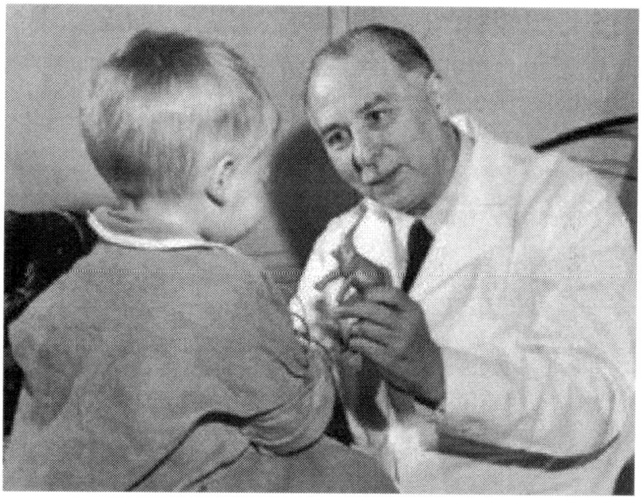

Sidney Farber: Dr. Sidney Farber (1903-1973) interacting with a young patient. [Dana-Farber Cancer Institute. Photo History: The Legacy of Dr. Sidney Farber. Photograph. September 30, 2014. https://blog.dana-farber.org/insight/2013/11/learning-from-history/]

Sidney Farber: Dr. Sidney Farber conducting research with coworkers in basement laboratory. [Dana-Farber Cancer Institute. Photo History: The Legacy of Dr. Sidney Farber. Photograph. September 30, 2014. https://blog.dana-farber.org/insight/2013/11/learning-from-history/]

of Chicago and the Sloane-Kettering Institute to repeat these trials with similar outcomes.[16]

Though World War II never saw the use of these sulfurous compounds, in 1943 the accidental exposure of troops to mustard gasses due to the Bari incident off the coast of Italy, and the autopsies that followed, revealed an important observation.[17] Every autopsied body showed the same suppression of the bone marrow and lymphatic tissue in a way that could only have occurred after exposure to these mustard gases.[18] While early investigations were done in utter secrecy due to the use of mustard gases as chemical warfare agents, by 1946, all information and research that had been done by Goodman and Gilman would be made public and nitrogen mustards would become generally available to everyone.[19]

Continuing the work of Goodman and Gilman, Doctor Sydney Farber was another cancer researcher hailing from Harvard Medical School and Boston Children's Hospital. Now considered the "Father of Chemotherapy," Farber was a prominent pediatric pathologist.[20] Based on earlier works discussing the importance of folic acid for bone marrow function and the recognition that folic acid stimulated cell-growth in leukemia patients, Farber "hypothesised [sic] that folic antagonists would inhibit or arrest the proliferation of cancer cells."[21] With virtually no cytotoxic agents available for the treatment of acute pediatric leukemias and lymphomas, they were inevitably fatal, yet despite the cynicism of the medical community, Farber made it his life's mission to improve the care and survival rates of children with cancer. With the development of folic acid inhibitors like aminopterin, now known as methotrexate, Farber would conduct his first clinical trials in 1947.[22]

Turning to folic acid inhibitors, Doctor Farber based his hypothesis on the results of over 200 post-mortem studies of children whose acute leukemia was treated by folic acid conjugates.[23] Of the children studied, there was a notable increase in bone marrow and abdominal cavity affectation in the children treated with folic acid conjugates. In comparison, those who were not treated in the same manner did not experience the same acceleration of cancer growth. Thus, Farber determined that it was worth exploring whether this acceleration could be of therapeutic benefit as a precursor to using radiation or nitrogen mustards.[24]

Following this line of thought, Farber justified the decision to pursue research toward folic acid antagonists based on the results of a 4-year-old girl with myelogenous leukemia that was rapidly progressing. From February 17 to March 24, 1947, this child was treated with a 100 to 300mg dose of diopterin (a folic acid conjugate) with little effect.[25] A bone marrow biopsy on March 25, 1947, showed continued bone marrow involvement. Following these results, Farber changed course and began administering 40mg of pteroyl aspartic acid (folic acid inhibitor) daily from March 28 to April 4, 1947. A post-mortem examination on April 4 showed virtually no bone marrow involvement, and a change of that mag-

Sidney Farber: Dr. Farber meeting with President Lyndon B. Johnson in an effort to lobby for government funding of cancer research. [Dana-Farber Cancer Institute. Photo History: The Legacy of Dr. Sidney Farber. Photograph. September 30, 2014. https://blog.dana-farber.org/insight/2013/11/learning-from-history/]

nitude as the result of chemotherapeutic drugs had not yet been seen during cancer research.[26]

Given the encouraging results of the previously mentioned child, Farber and his colleagues moved forward with their research. This time using 4-aminopteroyl-glutamic acid (aminopterin), Farber et al. began further study on another 16 patients.[27] Of these patients, six did not respond to treatment, and when Farber's study was published four had died and the other two remained unimproved. Regarding the other ten, the five children with the most promising results were reported, with side effects and the temporary nature of their remissions being stressed.[28] The remaining five patients were children ranging from 2 to 7 years old. Most of these children initially presented with fever, poor appetite, anemia, and enlarged lymph nodes, spleen, and liver. During their treatment, folic acid conjugates and folic acid inhibitors were alternated, with folic acid conjugates causing rapid decline in the children being studied. Symptoms worsened, side effects worsened, and all children saw significant increases in white blood cell count. When the folic acid inhibitor was begun, particularly aminopterin, all cases saw marked improvement. White blood cell counts decreased rapidly, appetites came back, and one child began walking again after two months in a wheelchair.[29] Despite many hospitalizations and setbacks, Case 4 was of a 2-year-old girl with the most promising results. Though remission was temporary for all involved patients, she went 43 days without treatment before needing to resume aminopterin. During her final round of chemotherapy with Doctor Farber, a second drug, teropterin, was added in combination with the aminopterin and this combination also gave very promising results.[30]

During Farber's study, numerous toxic side effects presented themselves in the children. Most common was reduced appetite and stomatitis with ulcers that typically healed and resolved themselves quickly upon cessation of treatment.[31] Other more severe side effects included infections and sepsis needing treatment via antibiotics, and severe anemia requiring blood transfusions. Later side effects would also be noted, these would include nausea, vomiting, hair loss, and neuropathy (pain caused by nerve damage).[32] Even amid poor reactions and negative side effects, further research led to other positive results. Published in 1953, Stefanini and Dameshek's work to treat excessive bleeding resulting from severe anemia led to the safe harvesting, storage, and transfusion of platelets.[33] This would do more than just help cancer patients—it would once again improve the safety of surgical procedures because there was now another way to control excessive bleeding.

While the results of Farber's study encouraged further research and clinical trials, concerns of over toxicity made developing other effective, but gentler, chemotherapeutic agents a pressing need.[34] Aminopterin later became known as methotrexate, one of the most well-known and frequently use chemotherapeutic agents on the market.[35] The success of Doctor Farber "ushered in the era of antimetabolite chemotherapy and generated great interest in the antifolate class of agents."[36] In the time since Farber conducted his studies, methotrexate has expanded in use for both cancerous and non-cancerous treatments. Additional diseases of note are "rheumatoid arthritis, psoriasis, and bacterial, fungal, and parasitic infections."[37] In the treatment of cancer, however, methotrexate has become a mainstay in the treatment of lymphomas and leukemias with cerebrospinal fluid and brain involvement.[38] Often administered intrathecally via lumbar puncture and spinal tap, methotrexate is a first line chemotherapeutic for both treatment and preventing the spread of cancer cells to the brain or spinal fluid.[39]

Chemotherapy and cancer treatment is an ever-evolving field of medicine. The first in a long line of chemotherapeutic agents, antifolate drugs are among the best understood classes of drugs, however, more have developed in the 70 years since the work of Gilman, Goodman, and Farber. Today, several options are available to patients meant to increase the efficacy of chemotherapeutics and reduce side effects.[40] Combination therapies, targeted gene therapies, hormonal treatments, immunological treatments, molecularly targeted therapies and anti-nausea medications are among the many treatment options meant to combat both cancer and many of the side effects allowing patients a better quality of life while undergoing treatment.[41] Improvements in both surgery and radiation have allowed for a nearly infinite number of treatment combinations showing great success.

With such groundbreaking and rapid success in cancer treatment since Goodman, Gilman, and Farber, the research of today is turning to cancer prevention rather than treatment. On October 5, 2023, a study regarding the life cycle of stem cells and their pre-cancer evolution was published in the *New England Journal of Medicine.* When determining what the future of medicine and cancer research holds, stem-cells seem to be the answer for many. With targeted gene therapy, providers can better determine the risk factors for certain patients and the idea is to alter the DNA of stem-cells before they can become anything malignant in a patient.[42] A stem-cell is nothing until it becomes something, so the prevention of cancer before it begins could solve many concerns regarding over toxicity resulting from the use of radiation and chemotherapeutics.

As pioneers in their respective fields, Goodman, Gilman, and Farber fundamentally reshaped both medicine and history. Revolutionizing a field that had remained stagnant for millennia, their groundbreaking contributions propelled humanity forward at an unparalleled pace. Despite encountering resistance from skeptics within the scientific community, Goodman, Gilman, and Farber persisted through obstacles, setbacks, and groundbreaking discoveries, ultimately achieving what many deemed unattainable.

Endnotes

1 Siddhartha Mukherjee. *The Emperor of All Maladies* (New York: Simon and Schuster, 2011), 39-41.

2 Ibid., 39-41.

3 Henry James Breasted, *The Edwin Smith Surgical Papyrus* (Chicago: The Chicago University Press 1980), 363-365.

4 Ibid., 407.

5 American Cancer Society, "Understanding What Cancer Is: Ancient Times to Present." Accessed October 20, 2023.

6 Johnson, A., and H. L. Lombard. "The Estimation of Operative Risk in Patients with Cancer." *The New England Journal of Medicine 224, no.* 18 (1941), 759.

7 Mukherjee, *Emperor of All Maladies,* 73-74.

8 Ibid., 75-76.

9 American Cancer Society, "History of Cancer Treatments: Radiation Therapy." Accessed October 20, 2023.

10 Ibid.

11 Vincent T. DeVita Jr., and Edward Chu. "A History of Cancer Chemotherapy." *Cancer Research 68*, no. 21 (November 2008): 8643-8644.

12 Goodman, Louis Sanford, and Alfred Gilman, *The Pharmacological Basis of Therapeutics* (The McMillan Company, 1956), 1415.

13 DeVita and Chu, "A History of Cancer Chemotherapy," 8645.

14 Goodman and Gilman, *The Pharmacological Basis of Therapeutics,* 1415.

15 Ibid., 1416.

16 Jennet Conant. *The Great Secret: The Classified World War II Disaster that Launched the War on Cancer,* 209, 260.

17 Ibid., 38.

18 DeVita and Chu, "A History of Cancer Chemotherapy," 8644.

19 Ibid., 1416.

20 Dennis R. Miller, "A Tribute to Sidney Farber the Father of Modern Chemotherapy." *British Journal of Haematology* 134, no. 1 (July 2006), 20.

21 Ibid., 20.

22 DeVita and Chu. "Cancer Chemotherapy," 8645-8646.

23 Sydney Farber, Louis K. Diamond, Robert D. Mercer, Robert F. Sylvester, and James A. Wolff. "Temporary Remissions in Acute Leukemia in Children Produced by Folic Acid Antagonist, 4-Aminopteroyl-Glutamic Acid (Aminopterin)," 787.

24 Ibid., 788-789.

25 Ibid., 789.

26 Ibid., 790.

27 Ibid., 790.

28 Ibid., 791.

29 Ibid., 791-793.

30 Ibid., 790.

31 Ibid., 793.

32 Kulmira Nurgali, R. Thomas Jagoe, and Raquel Abalo. "Editorial: Adverse Effects of

Cancer Chemotherapy: Anything New to Improve Tolerance and Reduce Sequelae." *Frontiers in Pharmacology* 9 (2018): 245.

33 Mario Stefanini, and William Dameshek. "Collection, Preservation and Transfusion of Platelets - With Special Reference to the Factors Affecting the Survival Rate and the Clinical Effectiveness of Transfused Platelets." *New England Journal of Medicine* 248, no. 19 (May 1953): 797.

34 Farber, et al., "Temporary Remissions," 793.

35 Bruce A. Chabner, and Dan L Longo. *Cancer Chemotherapy and Biotherapy: Principles and Practice*. 5th ed. Philadelphia: Wolters Kluwer Health, 2010: 109.

36 Chabner and Longo, *Cancer Chemotherapy*, 109.

37 Ibid., 109.

38 Ibid., 109.

39 Karen Arnold-Korzeniowski, "Intrathecal Chemotherapy (IT Chemo)."

40 American Cancer Society. "Chemotherapy."

41 Chabner and Longo, *Cancer Chemotherapy*, vii.

42 Catriona H.M. Jamieson, and Irving L. Weissman. "Stem-cell Aging and Pathways to Precancer Evolution." *New England Journal of Medicine* 389, no. 14 (October 2023): 1310-1312.

Bibliography

American Cancer Society. "History of Cancer Treatments: Chemotherapy." Accessed October 25, 2023. History of Cancer Treatments: Chemotherapy | American Cancer Society

American Cancer Society. "History of Cancer Treatments: Radiation Therapy." Accessed October 20, 2023. History of Cancer Treatments: Radiation Therapy | American Cancer Society

American Cancer Society. "Understanding What Cancer Is: Ancient Times to Present." Accessed October 20, 2023. Understanding What Cancer Is: Ancient Times to Present | American Cancer Society

Arnold-Korzeniowski, Karen. "Intrathecal Chemotherapy (IT Chemo)." Last reviewed November 17, 2021. Intrathecal Chemotherapy (IT Chemo) | OncoLink

Breasted, James Henry. *The Edwin Smith Surgical Papyrus*. Chicago: The University of Chicago Press, 1980: 363-390, 403-414. oip3.pdf (uchicago.edu)

Chabner, Bruce A, and Dan L Longo. *Cancer Chemotherapy and Biotherapy: Principles and Practice*. 5th ed. Philadelphia: Wolters Kluwer Health, 2010: 109-125. ProQuest Ebook Central - Reader

Conant, Jennet. *The Great Secret: The Classified World War II Disaster that Launched the War on Cancer*. New York: W.W. Norton & Company, Inc., 2020.

DeVita, Vincent T. Jr, and Edward Chu. "A History of Cancer Chemotherapy." *Cancer Research* 68, no. 21 (November 2008): 8643-8653. https://doi.org/10.1158/0008-5472.CAN-07-6611

Farber, Sidney, Louis K. Diamond, Robert D. Mercer, Robert F. Sylvester, and James A. Wolff. "Temporary Remissions in Acute Leukemia in Children Produced by Folic Acid Antagonist, 4-Aminopteroyl-Glutamic Acid (Aminopterin)." *New England Journal of Medicine* 238, no. 23 (June 1948): 787-93. https://doi.org/10.1056/nejm194806032382301.

Goodman, Louis Sanford, and Alfred Gilman. *The Pharmacological Basis of Therapeutics* (The McMillan Company, 1956), 1415-1450. The pharmacological basis of therapeutics; a textbook of pharmacology, toxicology and therapeutics for physicians and medical students : Goodman, Louis Sanford, 1906- : Free Download, Borrow, and Streaming : Internet Archive.

Jamieson, Catriona H.M., and Irving L. Weissman. "Stem-cell Aging and Pathways to Precancer Evolution." *New England Journal of Medicine* 389, no. 14 (October 2023): 1310-1319. Stem-Cell Aging and Pathways to Precancer Evolution | NEJM.

Johnson, A., and H. L. Lombard. "The Estimation of Operative Risk in Patients with Cancer." *The New England Journal of Medicine 224, no.* 18 (1941): 759-762. https://doi.org/10.1056/nejm194105012241802\

Miller, Dennis R. "A Tribute to Sidney Farber the Father of Modern Chemotherapy." *British Journal of Haematology* 134, no. 1 (July 2006): 20-26. https://doi.org/10.1111/j.1365-2141.2006.06119.x.

Mukherjee, Siddhartha. *The Emperor of All Maladies*. New York: Simon and Schuster, 2011.

Nurgali, Kulmira, R. Thomas Jagoe, and Raquel Abalo. "Editorial: Adverse Effects of Cancer Chemotherapy: Anything New to Improve Tolerance and Reduce Sequelae." *Frontiers in Pharmacology* 9 (2018): 245. https://doi.org/10.3389/fphar.2018.00245.

Stefanini, Mario, and William Dameshek. "Collection, Preservation and Transfusion of Platelets - With Special Reference to the Factors Affecting the Survival Rate and the Clinical Effectiveness of Transfused Platelets." *New England Journal of Medicine* 248, no. 19 (May 1953): 797-802. Collection, Preservation and Transfusion of Platelets — With Special Reference to the Factors Affecting the Survival Rate and the Clinical Effectiveness of Transfused Platelets | NEJM.

Moore v Dempsey: How a Pogrom Became a Turning Point in the Fight for Civil Rights

Gilda Bour

Independent Historian

Abstract

From the period of April to November 1919, the occurrence of over 25 large-scale racially charged riots in the United States, in which whites perpetuated violence against blacks in multiple cities, left thousands injured and hundreds of people dead. One of the worst of these events in Elaine, Arkansas, turned into a pogrom, where approximately 200 people were killed. In the aftermath, hundreds of black men were arrested, and many were indicted for murder. The case of *Moore v Dempsey*, resulting from the Elaine, Arkansas pogrom, served as a turning point for the protection of human and legal rights and led to a defining shift to the era with more concentration on civil rights. When the Supreme Court agreed in a 6-2 decision in 1923 that the interference of the mob violated the defendant's rights to due process, it negated the prior precedence of no review of the fairness of state criminal trials on petitions for writs of habeas corpus. At this juncture, the law became a crucial resource in the Civil Rights Movement and the fight for civil rights, and it was a vital turning point the NAACP would continue to build upon in the fight for civil rights and liberties. The racial riots of 1919 created awareness of the racial inequality that existed and the denial of democracy for many, and the case of Moore v Dempsey opened the door to more cases that would bolster the Civil Rights Movement of the 20th century.

Keywords: Elaine Massacre, Pogrom, NAACP, Elaine Twelve, *Moore v Dempsey*, Mob Violence, Moorefield Storey, Scipio Jones, Racial Riots 1919, due process

doi: 10.18278/sshj.12.1.6

Moore contra Dempsey: cómo un pogromo se convirtió en un punto de inflexión en la lucha por los derechos civiles

Resumen

Entre abril y noviembre de 1919, se produjeron más de 25 disturbios a gran escala con carga racial en los Estados Unidos, en los que blancos perpetuaron la violencia contra negros en varias ciudades, y dejaron miles de heridos y cientos de muertos. Uno de los peores acontecimientos en Elaine, Arkansas, se convirtió en un pogromo, donde murieron aproximadamente 200 personas. Posteriormente, cientos de hombres negros fueron arrestados y muchos fueron acusados de asesinato. El caso Moore contra Dempsey, resultante del pogromo de Elaine, Arkansas, sirvió como un punto de inflexión para la protección de los derechos humanos y legales y condujo a un cambio decisivo hacia una era con mayor concentración en los derechos civiles. Cuando la Corte Suprema acordó en una decisión de 6 a 2 en 1923 que la interferencia de la mafia violaba los derechos del acusado al debido proceso, negó el precedente de no revisión de la imparcialidad de los juicios penales estatales en peticiones de recursos de hábeas corpus. En esta coyuntura, la ley se convirtió en un recurso crucial en el movimiento de derechos civiles y la lucha por los derechos civiles, y fue un punto de inflexión vital que la NAACP continuaría aprovechando en la lucha por los derechos y libertades civiles. Los disturbios raciales de 1919 crearon conciencia de la desigualdad racial que existía y la negación de la democracia para muchos, y el caso Moore v Dempsey abrió la puerta a más casos que reforzarían el Movimiento de Derechos Civiles del siglo XX.

Palabras clave: masacre de Elaine, pogromo, NAACP, Elaine Twelve, Moore v Dempsey, violencia mafiosa, Moorefield Storey, Scipio Jones, disturbios raciales de 1919, debido proceso

摩尔诉邓普西案：大屠杀如何成为民权斗争的转折点

摘要

1919年4月至11月期间，美国发生了超过25起大规模种族骚乱，白人在多个城市对黑人实施暴力，造成数千人受伤、数百人死亡。其中最严重的事件之一发生在阿肯色州伊莱恩，最后演变成一场大屠杀，造成约200人死亡。随后，数百名黑人被捕，许多人被指控参与谋杀。阿肯色州伊莱恩大屠杀引发的摩尔诉邓普西案是“保护人权和合法权利”的转折点，并导致了向“更加注重公民权利的时代”的决定性转变。当最高法院于1923年以6比2的裁决认为，暴民的干涉侵犯了被告的正当程序权利时，它否定了“不对人身保护令申请进行国家刑事审判公正性审查”的先例。在这一关键时刻，法律成为了民权运动和争取公民权利的重要资源，也是美国全国有色人种协进会继续争取公民权利和自由的重要转折点。1919年的种族骚乱让人们意识到种族不平等以及对多数民主的否定，而摩尔诉邓普西案为更多案件打开了大门，这些案件将支持20世纪的民权运动。

关键词：伊莱恩大屠杀，大屠杀，美国全国有色人种协进会，伊莱恩十二名被告(Elaine Twelve)，摩尔诉邓普西案，暴徒暴力，Moorefield Storey，Scipio Jones，1919年种族骚乱，正当程序

From April through November 1919, racial unrest resulted in one of the worst instances of violent riots throughout the Midwest, South, Northeastern, and Great Plains parts of the United States. At the end of the summer of 1919, a planned union meeting in Arkansas turned into one of the deadliest massacres in United States history, described by some as a pogrom. More than 460 black men and women were arrested in two separate locations, and 122 were indicted for insurrection-related charges, and 73 for murder.[1] Twelve men were sentenced to death for their alleged involvement in the massacre, and the NAACP worked to fund and defend them. The resulting Supreme Court case of *Moore v Dempsey* that stemmed from defense of the twelve men provided a federal focus on the fairness of state criminal trials on petitions for writs of *habeas corpus*. The bloody riots, targeting black people and the massacre in Elaine, Arkansas, awakened the nation to the racial inequality that existed, as black men and women stood up to protect themselves and demand equal rights and the resulting Supreme Court case of *Moore v Dempsey* became a crucial turning point in the fight for Civil Rights in the 20th century.

As early as 1915, black people mobilized and migrated North and West, seeking better economic opportunities and improved social and housing conditions. One result is that they filled many positions left vacant by men fighting in World War I. This gave them more economic opportunities and improved wages. When the war ended abruptly in 1918, soldiers wanted to come home to a normal life like before the war. Both white and black men flooded major cities and began competing for jobs in a contracting labor market. Tensions increased as white veterans seeking work found many of their previously held positions had been taken by black laborers.[2] Rising prices for consumer goods, rents, and labor strikes also contributed to the escalating unrest.

Much of the white population was uneasy about change in the social, political, and economic order. Racially incited mob violence against blacks led to lynchings and riots as many blacks stood up to injustices and fought to gain rights long denied to them. Black veterans had fought to preserve the ideals of democracy abroad, and they were not willing to return home as second-class citizens to a homeland that denied them rights and freedoms. The black population that had migrated to find better housing and job opportunities was reluctant to surrender and return to the status quo before the war.

Across the nation, in response to the violence, blacks resisted and armed to defend themselves against white-initiated violence.[3] In the areas where the migrant population of blacks was largest, violent clashes turned into deadly and destructive riots. The worst of these occurred in Elaine, Arkansas, Chicago, Illinois, Omaha, Nebraska, and Washington, D.C. Over forty racially charged riots occurred across the nation throughout 1919, in which thousands were injured and hundreds died. The riots exposed a profound system of inequality and segregation, and black people resisted the status quo to gain economic and social prosperity and obtain their civil rights.[4]

At the end of summer in September 1919, black farmers near the town of Elaine in Phillips County, Arkansas, attended a Progressive Farmers and Household Union of America union meeting. While many blacks had migrated away from the South, Phillips County, Arkansas, continued to have a large black population in 1919. There was a robust middle class, with many blacks occupying positions like doctors, teachers, lawyers, clergy members, and business owners. Many of the black farmers worked large pieces of land as sharecroppers. During the war, cotton prices had risen due to the demand for supplies. They wanted to obtain the money owed, made from their share of the crops. They also wanted an itemized accounting for the goods charged from the local store. In the economic system of sharecropping, landowners, business owners, or merchants set the prices for cotton and other crops, didn't always provide an accounting of supplies given or loaned, and often charged enormous sums for settling accounts at season's end. Sometimes, sharecroppers would be in debt rather than break even or make money due to the cost of the supplies and goods. It was a system known as debt peonage.[5]

Robert Hill, a black veteran of WWI, was one of the original organizers of the black member union, the Progressive Farmers and Household Union of America, in early 1918.[6] Another union member was Ed Ware, a farmer who owned his land and had found some semblance of success. He joined the Hoop Spur lodge when it was newly formed in early August of 1919 and became the secretary.[7] Robert Hill worked with a white attorney named Ulysses S. Bratton, a renowned Republican lawyer in Little Rock, Arkansas, who also served as District Attorney and Postmaster of Little Rock.[8] He had been working with sharecroppers from the Arkansas Delta for about a year and aimed to break the system of debt peonage.[9]

On September 30, 1919, the Progressive Farmers and Household Union of America meeting was to occur in the evening at a country church in Hoop Spur. In attendance at the meeting was Ulysses Bratton's son, Ocier. In the early evening, Ed Ware drove to the church with approximately 100-200 men, women, and children.[10] When he entered the church, he quietly said the union password, "We've just begun."[11] Due to the number of threats the union had received, armed men were guarding the church, and the farmers brought shotguns and rifles to the meeting.[12] Robert Hill had not shown up to the meeting.

At about 11 p.m. that night, a car had driven up, stopped about 40 yards away, and shut off its lights and engine.[13] Two white men, Charles Pratt, a county deputy, and Will Adkins, a railroad detective, and a black prisoner who assisted local law enforcement called Kidd Collins, were in the car.[14] While Collins was a prisoner serving time for murder, the deputy gave him a revolver to carry for the trip.[15] What happened next was disputed, as there are two different accounts. According to Pratt, they were out looking for a white bootlegger, and it was only by chance the men stopped near the church.[16] When they stopped to urinate on the side of the road, black men approached them and opened fire.

According to the black men outside the church, Lit Simmons and John Ratliff, the car idled for approximately five minutes on the side of the road before three men got out of the vehicle.[17] One of the white men from the car flashed light from a flashlight and rapidly fired three shots from a pistol at the church. From there, someone unleashed from the back of the vehicle a torrent of bullets. Gunfire also erupted from the men guarding the church. Charles Pratt was hit in the knee and wounded, and Will Adkins, shot in the abdomen, died almost immediately.[18] Inside the church, people blew out the lights and dove to safety as pamphlets flew and glass from the windows littered the ground. The members inside the church also stated the men in the car fired first.[19]

After the shooting ended, over a hundred men, women, and children fled for safety. Most ran South of the church, while others hid in the bushes off the side of the road to see if more cars were coming. No blacks were seriously injured or killed that night. Within a half-hour, witnesses arrived to find Will Adkins's corpse and found the car full of bullet holes. Other witnesses that came that night said the church had no bullet holes, but other black and white witnesses later said the church was shot up, the inside in disarray, and windows destroyed.[20]

FIRE ON BLACKS
KILLING ONE

Military Patrol Fire on Negroes
Near Elaine, Ark.

Headline of a newspaper article from October 3, 1919. ["FIRE ON BLACKS KILLING ONE." Perth Amboy Evening News, October 3, 1919.]

Rumors of a black insurrection were spreading fast throughout the town of Elaine, and surrounding areas received news of unrest through articles they read. Newspapers ran stories of a negro plot to kill white people, and by the next day, had spread considerably. One story titled "Negroes Planned Revolt" supported this idea with the opening statement: "The trouble began with the discovery last night of an alleged negro plot to rise against the white residents of the southern part of Phillips county when Deputy Pratt with Adkins and a negro trusty were ambushed opposite a negro church at Hoop Spur, two miles north of Elaine, while on their way to arrest members of the Clem family...."[21]

The next day, the Sheriff of Phillips County and the Mayor of Helena ordered residents to stay inside and close local businesses and venues. Reports of phone lines being cut, with all communication with Elaine severed, were published in newspapers.[22] Descriptions of the white population outnumbered by the black population also were reported.[23] Calls for assistance from surrounding areas to help hunt down the men who shot Pratt and Adkins were sent out. This call brought men from Mississippi and Tennessee, and those in Arkansas and the Helena American Legion requested members to provide armed assistance.[24]

Early Wednesday morning after the initial shooting, Frank Kitchens, a plantation owner well known for his intense dislike of black people and Sherriff of Phillips County, showed up, ill, to the courthouse to direct affairs. He appointed Sebastian Staub as the new acting sheriff, appointed Herbert Thompson, a war veteran, as second in command, and deputized men.[25] Many of the few hundred deputized men who made up the posse were WWI veterans.

Convinced a black insurrection was in progress, Frank Kitchens divided the men into groups, with orders to subdue, disarm, and detain black people.[26]

Most of the black people who had fled the union meeting the night before weren't aware a man had died outside the church. Those who knew knocked on the doors of fellow union members and urged them to meet centrally at Frank Moore's home. Approximately 50 men showed up, armed with pistols and shotguns.[27] In the morning hours, the men gathered in the house heard shots fired from Hoop Spur. A small black boy ran down the road to the house and alerted them of the white mobs roaming the county. As a woman from the house, Nina Jenkins testified, the boy said, "He told us some white people were coming and said they were going to kill everything that was big enough to die."[28]

As blacks rushed about for safety and tried to understand what was happening, so did Ed Ware. He met up with other sharecroppers at a local cabin of Sallie Gile to discuss the intentions of the white mob.[29] Ed Ware looked out the cabin windows and saw white men walking, approximately four and five in a row, carrying guns and led by Kid Collins. They shouted to the inhabitants to come out of the cabin. Ed Ware's wife, Lulu, stepped outside to confront the approaching white men, and Ed left

Black men taken prisoner by soldiers in the Elaine Massacre; 1919. [Butler Center for Arkansas Studies, Bobby L. Roberts Library of Arkansas History and Art, Central Arkansas Library System]

to get his gun. Another man, Charlie Robinson, who'd been watching from his cabin, also tried to flee but couldn't do more than limp, and the white men shot him dead. All the other inhabitants of Sallie Gile's cabin escaped, except for Lulu, who was arrested and imprisoned in the local jail.[30]

Many of those fleeing the cabin ran to the brush near a marsh and another place half a mile away to the Hoop Spur union president, Jim Miller's home. As a mob of white men showed, the men met up and hid in the bushes. Many black men were armed but chose to hide in a wet marsh instead of fighting, hoping the white mob would pass. Instead, as the white men fired at the bushes above them, the black men realized the white men were shooting the women.[31] The white mob included the posse led by Herbert Thompson and another from Elaine, and both groups flanked the marsh, trying to flush the men out of hiding. Black men were immediately shot to death when they came into sight and not arrested. Many of those killed were unarmed. As the white posse fired upon the black men, some of the black men hiding in the marsh returned fire. Jack Tappan, a white veteran, was killed in the gunfire.[32] Upon hearing the gunfire exchange, additional black union men arrived to assist and saw multiple cars and the white mob of men. More gunfire erupted as both the black and white men shot at each other.

A white man sitting in one of the cars, Clinton Lee, was caught in the crossfire.[33] Word of the death of both Lee and Tappan spread rapidly through the county and incensed the white population and posses, and the white mobs were now out for both blood and vengeance. That evening, even though approximately 50 black men and one white man were jailed, the killing continued. Many of the white women and children had been put on trains earlier in the day and evacuated.[34]

Business and local political leaders sent telegrams to the Arkansas Governor, Charles Brough, seeking help. The Governor responded to the calls for assistance and escalated it further instead of trying to calm the situation. He called the United States Secretary of War, Newton Baker, and asked for troops from Camp Pike to assist with the problem in Phillips County.[35] After the telephone call, he sent a telegram to Baker that read, "RACE RIOT AT ELAINE PHILLIPS COUNTY THIS STATE FOUR WHITE SAID TO BE KILLED NEGROES SAID TO BE MASSING FOR ATTACK REQUEST COMMANDING GENERAL CAMP PIKE BE AUTHORIZED TO SEND SUCH UNITED STATES TROOPS AS MAY BE NECESSARY."[36] While Secretary Baker authorized Governor Brough's request, it was not until late that evening, after lengthy preparations, that the troops left Camp Pike to head to Phillips County.[37] Approximately 400 soldiers from the Third Division and 150 soldiers from the Fifth Infantry Division traveled from Little Rock by train.[38] Colonel Isaac Jenks commanded them.[39] Many of the soldiers were experienced combat veterans from World War I in France. All the soldiers were white, and their provisions included supplies of six trucks, two ambulances,

Right, Charles H. Brough talking with a U.S. Army officer in the aftermath of the Elaine Massacre' 1919. [Butler Center for Arkansas Studies, Bobby L. Roberts Library of Arkansas History and Art, Central Arkansas Library System]

twelve machine guns, and ample ammunition.[40] Wanting to witness the situation firsthand, Governor Brough was also on the train.

Arriving in Elaine the following day, Governor Brough and Colonel Jenks surveyed the area. One of the Captains that accompanied Brough and Jenks, Edward Passailiague, was met by a group of white civilians who explained how blacks had the town and farms surrounded and were shooting at any who tried to leave their farms or towns.[41] Colonel Jenks set up a command center from a telephone building in Elaine. Soldiers positioned a machine gun at the top of the building, and Jenks surveyed the scenes around town with binoculars.[42] The local town merchants gave shotguns, rifles, and guns to the soldiers, who gave them to the local posse.[43] The deputized men would accompany the soldiers to assist with disarming both black and white civilians, but the orders went further. The *Arkansas Democrat* reported, "The troops are all under orders to shoot to kill any negro who refuses surrender immediately."[44]

Colonel Jenks declared martial law, and shortly after 9:15 a.m., the troops marched through town and out into the fields.[45] As the soldiers marched along the road, many frightened black men, women, and children met them on the road, surprising the soldiers when they were unarmed. The soldiers put the blacks that met them on the road in the white schoolhouse and marched many more in gang lines led by the soldiers into town. Vacant buildings and the basement of the new school building were used to intern those captured.[46]

At the same time, Governor Brough, Colonel Jenks, and soldiers drove west of town, where black people were hiding, fearful. When four black men were frightened out of the bush, they shot at the car and soldiers. While the bullets went astray, it was established the black mob had attacked the Governor. Later that afternoon, a skirmish in the thick cane where one soldier was killed and another wounded would change how the soldiers cautiously pursued blacks.[47] The other soldiers opened fire into the thick cane brush and then worked their way through the thick brush with machine guns over the next hour. The reporter from the *Memphis Press* accompanying the soldiers said they saw many blacks being shot and that the soldiers used their machine guns on blacks.[48]

The posse and the troops worked in tandem, firing upon and rounding up blacks. Plantation owner Gerard Lambert witnessed the soldiers "mow down" a black man who ran from behind a building where he had been hiding.[49] Lambert also told how soldiers brought a black man as a prisoner to his store for interrogation and tied him to a column. The black man was then doused in kerosene, set on fire, and when he broke free of his bonds and ran, he was fired upon numerous times and shot dead.[50] Other soldiers spoke to witnesses and said that "this should be a lesson to them."[51] As the troops cleared fields and poked their rifles into the brush to find and flush out black men, those that were hiding fled deeper and further into the wood and brush, and the soldiers followed, firing. A soldier stated, "they were shooting them down like rabbits."[52] Governor Brough said later, "they took machine guns out there and let 'em have it.'[53]

Later in the afternoon, Colonel Jenks and his soldiers headed back to town with the prisoners they had captured. As they drove, they passed a plantation where about 50 white women and children were gathered and scared. Colonel Jenks ordered about a dozen soldiers to protect the white women and children against the blacks that had murderous plans.[54] Not far down the road, another event took place. Four prominent black men, brothers by the name of Johnston, had been on a hunting trip and had heard of the unrest.

Rather than drive, they had boarded a train to head back to Elaine. Two of the brothers, Leroy and Gibson, had recently returned from fighting in France, Louis was a doctor visiting from Oklahoma and D.A.E. was a dentist.[55] When their train arrived in Elaine, a group of white men was waiting to arrest them for their suspected part in the insurrection. The four brothers were taken from the train, chained together, and put in a car. They would later be found shot to death and dumped on the roadside near Elaine. It was unclear if the car they were riding in was shot up and the men killed or if one of them had tried to grab a gun and was then shot and killed. It was a stretch to attempt to connect them to the unrest occurring in Phillips county.

Once he returned to town, Colonel Jenks reported back to his superiors that the situation was under control, but it was not. The troops continued to spread out through town, firing upon and pursuing the black citizens.[56] As black men and women fled, the white mobs and soldiers ransacked their homes. From the black sharecropper's cabins, approximately 400 guns and 200 pistols were confiscated.[57] Gun ownership was not unusual, as it was customary for both black and white farmers to own weapons. As the posses and soldiers ransacked and searched and moved from cabin to farm, it caused the frightened black citizens to flee further and set up a cycle where the possess and soldiers continued to pursue them. A journalist, Sharpe Dunaway, wrote later the soldiers, working together with the posses, went from cabin to cabin, killing blacks.[58] For two days, the soldiers and posses roamed the countryside, pursuing and shooting blacks and generously used their machine guns.[59] Not many stories of the black citizens lying dead or being killed were reported in the newspapers. The few stories that did make it to publication gave hints as to the killing that occurred. The *Chattanooga News* published a story that reported, "The known negro dead today was fourteen, with other bodies reported in the cane brakes and underbrush about Elaine, where most of the fighting took place."[60]

Historian Charles L. Lumpkins defines pogrom as "an assault, condoned by officials, to destroy a community defined by ethnicity, race, or some other social identity."[61] In his book, journalist Robert Whitaker carefully details 22 separate and two anecdotal killing sites where over 200 blacks were reportedly killed.[62] Across these sites, the events that led to the racially targeted slaughter of these people were done by deputized men who made up posses and soldiers across four days. Newspapers across America proclaimed the event to be an uprising and a planned insurrection. Ida B. Wells detailed the event as a massacre.[63] It was more; it was a pogrom.

That Friday, Governor Brough traveled to Phillips County Courthouse and met community leaders there. He appointed a "Committee of Seven" from these leaders, and they were to research the planned insurrection and decide who should be prosecuted. All of these men were large landowners who employed many of the Phillips County sharecroppers. Governor Brough traveled back to Little Rock on October 4, 1919, and declared peace to be restored, yet the killing continued sporadically until Monday, October 6, 1919.[64]

The Committee of Seven had promised justice to the mob in the form of electrocuting those responsible. They were going to identify who was guilty, charge them with a crime, prosecute, and electrocute them. Yet organizing to obtain equal rights and fair prices for labor, while frowned upon by the county's white population, was not a crime punishable by electrocution. *The New York Times* ran a story detailing the planned slaughter of white people in Phillip's County, as told by the Committee of Seven on Monday, October 6, 1919.[65] The Committee obtained the details from forced and tortured confessions from blacks in custody. Robert Hill was also identified as the uprising leader and was being sought at the time of the article.[66]

Over 400 black men and women were taken into custody and held in a schoolhouse in Elaine, and approximately 60 were taken into custody in Helena.[67] Approximately 60 black men were held in custody in Helena. Only one white man was arrested, Ocier Bratton, the son of Ulysses Bratton, the attorney who had agreed to represent the union members.[68] He was charged with the murder of Will Adkins at the church in Hoop Spur and had to be smuggled out of town to prevent lynching. His father worked to have the charges dropped about a month later. Walter White, assistant secretary of the NAACP, traveled by train to perform an investigation on the incident and arrived in Phillips County on October 11, 1919.[69] He posed as an investigative reporter to get information and interviewed sharecroppers and Governor Brough.[70] After several more interviews and looking around, he felt threatened enough to flee the area.

Colonel Jenks interrogated the prisoners, as did the Committee of Seven. When the prisoners were interrogated, no attorneys were present.[71] For days, the prisoners were cruelly interrogated. An imprisoned black sharecropper was brought to the top floor, blindfolded, stripped, and made to lay on the concrete floor facedown.[72] Some were held down spread eagle and whipped, and others received electric shocks until confessions were received. T.K. Jones, one of the interrogators, said:

> They were not only whipped but formaldehyde was put to their noses and were stripped naked and put into an electric chair which they had in the room to frighten further and torture them. I not only personally saw a great many negroes whipped with a leather strap that would cut blood at every lick, but I whipped probably two dozen of them myself … one of the witnesses against the petitioners was whipped two or three times and put in the electric chair to make him testify against the petitioners.

A doctor had to attend to some of the prisoners because the torture had been so brutal.

Prisoners remained jailed without changes of clothing and proper food to eat. A total of fifty of the prisoners, including both men and all the women, were released, and the remaining men's prosecution moved quickly. John Miller served as the prosecutor. On October 27, a grand jury, including two members of the Committee of Seven, indicted seventy-three black men for first-degree murder and forty-nine for lesser charges.[73] The trials began the following Monday, November 3, 1919, and were fast. By the end of Tuesday, eleven men had been convicted of murder.[74] Ed Ware was arrested in New Orleans and brought back to stand trial for murder. He was also convicted and, along with the other eleven men, sentenced to die in the electric chair.[75] By Tuesday, November 11, 1919, all remaining men pleaded to lesser charges, such as second-degree murder, assault to kill, and night riding, to avoid the death penalty.[76]

Ulysses Bratton traveled to the NAACP's Manhattan office on October 30, 1919, to get assistance for the men. While the National Association for the Advancement of Colored People, or the NAACP, had never become involved in assisting criminal defendants, Bratton felt different.[77] He wanted the NAACP to help obtain counsel for the men in their appeals case and to put the case under a national review lens.[78] Over the next few weeks, Bratton met with the board members to convince them to support and defend the men.

Although the organization had started small, with limited funds and a few employees, it had quickly grown in the ten years before 1919. The

A picture of the twelve Elaine massacre defendants. [Butler Center for Arkansas Studies, Bobby L. Roberts Library of Arkansas History and Art, Central Arkansas Library System]"

NAACP's primary efforts were to create awareness about the mistreatment of blacks to highlight the inequality and work towards lobbying Congress for change and getting an anti-lynching bill passed.[79] They had never assisted criminal defendants before, either with the appeals process or trials in the state court. This practice changed as they launched a massive membership drive and fundraising campaign to assist the twelve men in Arkansas.[80] The NAACP legal defense fund was kept quiet, as there was the potential for harm if it were found out the NAACP was funding the legal defense of the twelve condemned men and might hurt their case in court.[81] The NAACP publicly campaigned to pressure Congress to launch an investigation into the Arkansas riot. Yet Congress and the Justice Department refused to get involved, would not make a statement, or publicly investigate what occurred in Arkansas.[82]

Bratton recommended a former Confederate colonel named George Murphy to represent the men.[83] Murphy was an established lawyer with a good reputation. At the same time, the NAACP had also been working with another law firm in Little Rock to potentially represent the men. Law partners Thomas Price and Scipio Jones had also reached out to the NAACP to represent the men.[84] They had already started working on the case of the 12 men.

Scipio Jones, Thomas Price, and Colonel George Murphy began working together by the end of November. Scipio Jones interviewed sharecroppers, gathered court records and transcripts, filed paperwork with the courts, and wrote briefs while George Murphy was the face of the defense team with the press and in the courtroom.[85] The NAACP was wary of having a black lawyer on the case and wanted to ensure that a white lawyer also represented the men.

Scipio Africanus Jones was born a slave towards the end of 1863, attended an all-black school as a child, and graduated from college, and became a schoolteacher in his twenties.[86] He applied to law school at the University of Arkansas but was turned down as expected, as there were no black students at the University.[87] Instead, Jones began to study law in the offices of a few local judges while teaching school. He passed the bar exam on June 15, 1883 and began practicing law in the Arkansas inferior and circuit courts.[88] As Scipio Jones was working towards becoming a prominent black lawyer, Jim Crow laws began to appear, along with the disenfranchisement of blacks. Despite this, Jones prospered, becoming a landowner and attorney for a renowned organization, the Mosaic Templars.[89] He worked hard to help the black community of Little Rock, where he lived and worked. Scipio was active in civic affairs and fought discriminatory laws.[90] In 1901, he became the first black lawyer in Arkansas to argue that his black client's due process had been violated due to the exclusion of blacks from the juries.[91] While he did not win this case, it would not be the last time he pursued this argument.

Scipio Africanus Jones, the attorney that fought for justice for the Elaine defendants. [Butler Center for Arkansas Studies, Bobby L. Roberts Library of Arkansas History and Art, Central Arkansas Library System]

In early December, Scipio Jones wrote a compelling argument for a motion to obtain a new trial for each of the 12 men sentenced to die. While the judge denied the motions, the 12 condemned men were granted 60 days with which they were able to prepare their appeals. When Scipio Jones was working on the appeals, the Bill of Rights did not limit the states regarding due process, and this had been repeatedly affirmed through many previous cases. Despite this, Scipio Jones was determined to fight for justice for his clients, and his hope for appeal rested on one case. In *Frank v Mangum* (1915), a Jewish man had been tried for the murder of a thirteen-year-old girl in Atlanta. Mob violence dominated the trial by those who wished to lynch him if he was not found guilty and sentenced to death.[92] Upon appeals, his guilty verdict was upheld by both the state and U.S. Supreme Courts. It was found that by allowing higher courts to review the lower courts' state criminal proceedings, Frank was provided due process.[93] Justice Holmes dissented and stated, "We are not speaking of mere disorder, or mere irregularities in procedure, but of a case where the processes of justice are actually subverted. In such a case, the Federal court has jurisdiction to issue the writ."[94]

On March 22, Scipio Jones and George Allen argued their appeals case in the Arkansas Supreme Court.[95] In the appeals decision, the Supreme Court of Arkansas ordered new trials for six men due to a "statutory law" flaw.[96] The other six men had their verdicts reaffirmed. This action split the group of twelve men into two distinct groups. The group that would receive new trials became known as the Ware defendants, and the six that had their convictions upheld and reaffirmed became known as the Moore defendants.[97]

The new trial for the Ware group got underway on May 3, 1920, in Helena, Arkansas.[98] The six defendants were again found guilty and sentenced to death on July 23, although Governor Brough granted an additional stay for an appeal.[99] In December of 1920, the Arkansas Supreme Court reversed Judge Jackson's decision on the Ware group of men on discriminatory grounds for not having any blacks on the petit or grand jury.[100] Governor Brough's term in office had nearly expired, and when the new Governor of Arkansas took office, he set the execution date for the Moore group of men to June 10.[101]

Edgar McHaney and Jones worked fast to file a writ of *habeas corpus* in federal court when this occurred. In response, the Attorney General for Arkansas filed a writ of prohibition intending to carry out the executions of the Moore case men.[102] Oral arguments were instead scheduled when the state supreme court refused to overturn the writ of *habeas corpus*. At the hearing, both sides cited *Frank v Mangum* in their arguments. Scipio Jones and Ed McHaney argued that in the case of *Frank v Mangum*, there was precedent for a judge to find that a trial court had lost jurisdiction.[103] The Arkansas state assistant attorney general argued that according to the precedent set in *Frank v Mangum*, the appellate review had been completed, and no further argument could be made.[104] On June 20, 1921, the Arkansas Supreme Court ruled against the Moore men. A *New York Times* article reported, "Today's writ held that no State court of equity has jurisdiction over criminal cases and that the sentence of death stands."[105]

The next day, Judge J.M. Jackson of the First District Circuit Court granted a change of venue for the Ware men, to Lee County and not in Phillips County.[106] A few weeks later, Arkansas Governor McRae set the execution date for September 23, 1921, for the Moore group of men.[107] Judge Cotteral found probable cause for an appeal to the United States Supreme Court—a glimmer of hope.[108] The trial of the Ware men was set to begin on October 10, yet the prosecution asked for a continuance when two new witnesses for the defense were presented.[109] McHaney and Jones had secured two white men who provided affidavits of the torture that occurred to elicit the original confessions and also that James Tappan, the man the Ware group men were awaiting retrial for killing, was shot and killed by his own posse.[110]

The NAACP worked fast to hire a lawyer to assist with the argument of the case before the U.S. Supreme Court. Moorfield Storey, the former board president of the NAACP, was chosen.[111] He had argued twice before the Supreme Court. Scipio Jones put together a well-written habeas petition that laid out in detail the plight of the sharecroppers, and the narrative was one of white mob intimidation, torture, inequality, and a denial of due process.[112] In January of 1922, a few days before the Supreme Court case was to be argued, a decision was made to have Moorfield Storey and Ulysses Bratton argue before the court. The NAACP President, Walter White, wired Scipio Jones to tell him he would not be needed, even though he'd done most of the work.[113] On the afternoon of January 9, 1922, the case was argued by Bratton and Storey in front of Chief Justice William Taft's Supreme Court.[114] The courts were not known for ruling in favor of the civil rights of black men. Since the Fourteenth Amendment was enacted, it had pertained to black rights only 28 of 604 times as of 1912.[115] For an amendment intended to provide equal rights to black men, it was utilized more for big business. The civil rights cases that dealt with the rights of black men invoked the principle of state's rights.[116] Yet when it involved property rights for wealthy citizens, state's rights were not considered similarly. Taft's Supreme Court offered little hope that the ruling would differ in this case.

A few days before Storey argued in front of the Supreme Court, a race war broke out and was splashed across newspaper headlines. White mobs hunted and killed blacks, burned homes and businesses as the entire black section of town was torched in Rosewood, Florida.[117] "Thousands of persons were pouring into the village early this morning in heavy laden automobiles, all of them armed."[118] This event provided the opening backdrop for Storey to argue about how whites killed black men in Arkansas. It might not have been the legal point, but it helped paint the violent mob-dominated picture for the debated legal question. Bratton laid this out in his argument about the end of the summer of violence in Elaine, Arkansas, and the continuation of white instigated mob violence in both Tulsa and Rosewood.[119]

On February 19, 1923, the United States Supreme Court ruled in favor of the petitioners. In a 6-2 decision, the Court ruled the men's Constitutional right to due process was violated due to mob-dominated trials. Justice Holmes built upon his dissent in *Frank v Mangum* when he penned his opinion for *Moore v Dempsey*. Justice Holmes wrote,

> If the State, supplying no corrective process, carries into execution a judgment of death or imprisonment based upon a verdict thus produced by mob domination, the State deprives the accused of his life or liberty without due process of law. We assume in accordance with that case that the corrective process supplied by the State may be so adequate that interference by habeas corpus ought not to be allowed. It certainly is true that mere mistakes of law in the course of a trial are not to be corrected in that way. But if the case is that the whole proceeding is a mask—that counsel, jury and judge were swept to the fatal end by an irresistible

> wave of public passion, and that the State Courts failed to correct the wrong, neither perfection in the machinery for correction nor the possibility that the trial court and counsel saw no other way of avoiding an immediate outbreak of the mob can prevent this Court from securing to the petitioners their constitutional rights.[120]

It was a surprising ruling and one that struck a blow at Jim Crow. It was a great victory for the Arkansas men and the NAACP, that had fought hard to obtain justice. Mob intimidation and sham trials had been the status quo until this point, and there was no federal review of state proceedings. The Moore decision changed this and stepped away from the precedent in the *Frank v Mangum* case and became a constitutional victory that built a foundation for further constitutional victories in the future, such as victory in the case of *Brown v Board of Education* (1954). It signaled a willingness to review state court decisions in criminal matters.[121] The attorney that argued for Leo Frank in *Frank v Mangum* said of the Moore decision, "Due process of law now means, not merely a right to be heard before a court, but that it must be before a court that is not paralyzed by mob domination."[122]

The Moore defendants were not free, but the men were no longer facing the imminent threat of electrocution. The Ware defendants awaiting retrial had their cases dismissed as two terms of the court had passed without trials, and they had not been afforded a speedy trial as guaranteed by due process of law.[123] The Ware men were finally set free. Scipio Jones worked out a compromise for the remaining Moore men, and on January 13, 1924, Governor McRae granted the men "indefinite furloughs and set them free."[124]

By the beginning of the 20th century, as the nation retreated from Reconstruction, habeas corpus was disregarded as a means of government enforcement. Yet in the first half of the 20th century, three critical cases would help to change this and to "expand the causes and review procedures for federal habeas review of state due process violations."[125] *Frank v Mangum* cracked the door open for *Moore v Dempsey*, which further opened the door for the civil rights victory in *Brown v Board of Education*. *Moore v Dempsey* helped change the way due process violations were reviewed. This was no small feat, and each case built upon the other. The Moore case marked a significant point in history, where the Supreme Court showed a willingness to intervene in state criminal court proceedings that were not just.[126] The precedents set in previous cases that had repeatedly confirmed this were suddenly swept aside, and the foundation was laid for further individual and civil rights fights.

In November of 1919, the summer of racial violence ended, yet what occurred would lay the groundwork for the future. Black America was bolstered to continue to fight for equal rights. In the violence that erupted, many black Americans refused to relinquish the fight for equality, and many fought back to defend not only their homes, families, and lives but to obtain civil and individual rights. The NAACP learned and grew as an organization from the experiences they gained in defense of the condemned men from Arkansas. The case of *Moore v Dempsey* was a vital turning point and one the NAACP would continue to build upon in the fight for civil rights and liberties. The racial riots of 1919 that awoke a nation to the racial inequality that existed and the denial of democracy for many became a genesis of the Civil Rights Movement of the 20th century.

Endnotes

1 Cameron McWhirter, *Red Summer: The Summer of 1919 and the Awakening of Black America* (New York, NY: St. Martin's Griffin, 2012), 222-226.

2 McWhirter, 19.

3 David Frederick Krugler, *1919, The Year of Racial Violence: How African Americans Fought Back* (New York, NY: Cambridge University Press, 2015), 4.

4 Krugler, 5.

5 Krugler, 166.

6 McWhirter, 214.

7 McWhirter, 214.

8 Grif Stockley, Brian K. Mitchell, and Guy Lancaster, *BLOOD IN THEIR EYES: The Elaine Massacre of 1919* (Fayetteville, AR: University of Arkansas Press, 2020), 42.

9 Krugler, 166.

10 McWhirter, 216.

11 Robert Whitaker, *On the Laps of Gods: The Red Summer of 1919 and the Struggle for Justice That Remade a Nation* (New York, NY: Three Rivers Press, 2009), 2.

12 McWhirter, 216.

13 Whitaker, 14.

14 McWhirter, 216.

15 Whitaker, 81.

16 McWhirter, 216.

17 Whitaker, 83.

18 Krugler, 171.

19 McWhirter, 216.

20 Ibid., 217.

21 "Negroes Planned Revolt," *The Kansas City Times*, October 2, 1919, 1.

22 "Outbreak Reported at Elaine," *Arkansas Democrat*, October 1, 1919, 1.

23 Ibid.

24 McWhirter, 217.

25 Whitaker, 88.

26 Krugler, 172.

27 Whitaker, 89.

28 Ibid.

29 Ibid., 90.

30 Ibid.

31 Ibid., 91.

32 Krugler, 173.

33 Ibid.

34 Ibid., 97.

35 McWhirter, 219.

36 Ibid.

37 "New Race Battle," *The Kansas City Times*, October 2, 1919, 1.

38 McWhirter, 220.

39 Krugler, 176.

40 McWhirter, 220.

41 Whitaker, 108.

42 Ibid.

43 Ibid., 109.

44 "Ordered to Shoot to Kill," *Arkansas Democrat*, October 2, 1919, 1.

45 Whitaker, 109.

46 Ibid., 110.

47 McWhirter, 221.

48 Whitaker, 111.

49 Krugler, 177.

50 McWhirter, 224.

51 Krugler, 177.

52 Whitaker, 112.

53 Ibid.

54 Ibid., 113.

55 Ibid., 114.

56 McWhirter, 222.

57 Whitaker, 117.

58 Ibid., 118.

59 McWhirter, 222.

60 "Arkansas Wood Scene of Battle," *The Chattanooga News*, October 3, 1919, 1.

61 Charles L. Lumpkins, *American Pogrom: The East St. Louis Race Riot and Black Politics* (Athens, OH: Ohio University Press, 2008), xii.

62 Whitaker, 326-329.

63 Ida B. Wells-Barnett, *The Arkansas Race Riot* (Chicago, IL: Aquila, 1920), 51.

64 McWhirter, 223.

65 "Planned Massacre of Whites Today," *The New York Times*, October 6, 1919, 1.

66 Ibid.

67 McWhirter, 222.

68 "16 Are Dead in Race Riots In Arkansas," *The Journal*, October 3, 1919, 5.

69 Whitaker, 151.

70 Whitaker, 152.

71 McWhirter, 222.

72 Whitaker, 161.

73 Whitaker, 168.

74 Ibid., 179.

75 McWhirter, 227.

76 Whitaker, 180.

77 Whitaker, 202.

78 Ibid., 203.

79 Ibid., 202.

80 McWhirter, 228.

81 Ibid.

82 Whitaker, 203.

83 Stockley, Mitchell, and Lancaster, 181.

84 Ibid., 182.

85 Ibid., 205.

86 Tom Dillard, "Scipio A. Jones," *The Arkansas Historical Quarterly* 31, no. 3 (1972): 201-203.

87 Whitaker, 189.

88 Dillard, 204.

89 Whitaker, 190-191.

90 Dillard, 213.

91 Whitaker, 194.

92 Francis, 128.

93 Ibid.

94 Richard C. Cortner, *A Mob Intent on Death: The NAACP and the Arkansas Riot Cases* (Middletown, CT: Wesleyan University Press, 1988), 143.

95 Whitaker, 221.

96 Whitaker, 222.

97 Stockley, Mitchell, and Lancaster, 27.

98 Whitaker, 223.

99 Stockley, Mitchell, and Lancaster, 215.

100 Ibid., 103.

101 Stockley, Mitchell, and Lancaster, 226.

102 Stockley, Mitchell, and Lancaster, 232.

103 Whitaker, 256.

104 Ibid.

105 "Says Six Negroes Must Die," *The New York Times*, June 21, 1921, 27.

106 "Cases to Marianna," *Daily Arkansas Gazette*, June 21, 1921, 1.

107 "McRae Sets Sept. 23 for Execution of Elaine Blacks," *Pine Bluff Daily Graphic*, August 13, 1921, 1.

108 "State Sustained in Elaine Negroes' Case," *Arkansas Democrat*, September 27, 1921, 1.

109 Stockley, Mitchell, and Lancaster, 250.

110 Cortner, 121.

111 Ibid., 252.

112 Scipio Jones, "Scipio Jones' Brief to the Supreme Court Regarding the Elaine Twelve," (Little Rock, 2020), 2-5.

113 Stockley, Mitchell, and Lancaster, 257.

114 Ibid.

115 Whitaker, 273.

116 Ibid., 274.

117 "Whites and Negroes Slain in Rosewood Night Battle; Village Is Burned By Mob," *The Tampa Times*, January 5, 1923, 1.

118 Ibid.

119 Whitaker, 283.

120 "*MOORE Et Al. v DEMPSEY*, Keeper of Arkansas State Penitentiary," Legal Information Institute (Cornell Law School, 2020), https://www.law.cornell.edu/supreme-court/text/261/86, 1.

121 Francis, 129.

122 Ibid.

123 Stockley, Mitchell, and Lancaster, 263.

124 Whitaker, 305.

125 Justin J. Wert, *Habeas Corpus in America: The Politics of Individual Rights* (Lawrence, KS: University Press of Kansas, 2011), 117.

126 Francis, 128.

Bibliography

"16 Are Dead in Race Riots In Arkansas." *The Journal*. October 3, 1919.

"Arkansas Wood Scene of Battle." *The Chattanooga News*, October 3, 1919.

"Cases to Marianna." *Daily Arkansas Gazette*, June 21, 1921.

Cortner, Richard C. *A Mob Intent on Death: The NAACP and the Arkansas Riot Cases*. Middletown, CT: Wesleyan University Press, 1988.

Dilliard, Tom. "Scipio A. Jones." *The Arkansas Historical Quarterly* 31, no. 3 (1972): 201-19.

Jones, Scipio. "*Scipio Jones' Brief to the Supreme Court Regarding the Elaine Twelve.*" Little Rock: University of Arkansas Libraries, 2020.

Krugler, David Frederick. *1919, The Year of Racial Violence: How African Americans Fought Back*. New York, NY: Cambridge University Press, 2015.

Lumpkins, Charles L. *American Pogrom: The East St. Louis Race Riot and Black Politics*. Athens, OH: Ohio University Press, 2008.

"McRae Sets September 23 for Execution of Elaine Blacks." *Pine Bluff Daily Graphic*, August 13, 1921.

McWhirter, Cameron. *Red Summer: The Summer of 1919 and the Awakening of Black America*. New York, NY: St. Martin's Griffin, 2012.

"*MOORE Et Al. v DEMPSEY*, Keeper of Arkansas State Penitentiary." Legal Information Institute. Cornell Law School, 2020. https://www.law.cornell.edu/supreme court/text/261/86.

"Negroes Planned Revolt." *The Kansas City Times*, October 2, 1919.

"New Race Battle." *The Kansas City Times*, October 2, 1919.

"Ordered to Shoot to Kill." *Arkansas Democrat*, October 2, 1919.

"Outbreak Reported at Elaine." *Arkansas Democrat*, October 1, 1919.

"Planned Massacre of Whites Today." *The New York Times*, October 6, 1919.

"Says Six Negroes Must Die." *The New York Times*, June 21, 1921.

"State Sustained in Elaine Negroes' Case." *Arkansas Democrat*, September 27, 1921.

Stockley, Grif, Brian K. Mitchell, and Guy Lancaster. *BLOOD IN THEIR EYES: The Elaine Massacre of 1919*. Fayetteville, AR: University of Arkansas Press, 2020.

Wells-Barnett, Ida B. *The Arkansas Race Riot*. Chicago, IL: Aquila, 1920.

Whitaker, Robert. *On the Laps of Gods: The Red Summer of 1919 and the Struggle for Justice That Remade a Nation*. New York, NY: Three Rivers Press, 2009.

"Whites and Negroes Slain in Rosewood Night Battle; Village Is Burned By Mob." *The Tampa Times*, January 5, 1923.

Book Review: Nathan Prefer's *The Luzon Campaign 1945: MacArthur Returns*

Dr. Robert Young
Professor, Department of History and Military History
American Military University

Prefer, Nathan N. *The Luzon Campaign 1945: MacArthur Returns*, Philadelphia, Casemate Publishers, 2024. ISBN 978-1636244242. Notes. Index. Bibliography. Pp. 335. Hardcover $37.95. Electronic version available.

Nathan N. Prefer is a military history writer focusing on World War II. In this work, Prefer does not examine but rather renarrates the eight-month campaign for the Philippine island of Luzon. Aside from quick descriptions of individual achievements, such as a remarkable infantryman, artilleryman, or medic, this book fails to add anything of merit to the scholarship on this subject.

Individual heroes do receive praise within the pages of the book. It began in the introduction with Army Major Thomas Buchanan McGuire. A 24-year-old fighter pilot, multiple times an ace, the following account noted his bravery: "… more of McGuire's fellow pilots were being shot out of the skies. As he shot down his fourth enemy plane, his guns were emptied. Rather than leave the fight, Major McGuire remained and made 'dummy' passes at the enemy fighters to keep them away from the vulnerable bombers" (9). In Manila, he describes Private First Class Elbert E. Jones, a bazooka man, in the middle of a vicious fight: "… managed to snake his way up an additional 50 yards forward, dragging his weapon with him. Hiding behind the only cover, a dirt mound no higher than six inches tall, he fired 20 rounds of antitank ammunition directly into the station before return machine-gun fire killed him. His fire drove the Japanese from the left wing of the

 doi: 10.18278/sshj.12.1.7

station and kept the Japanese occupied while behind PFC Jones, medical aidmen carried off the wounded" (92). A final example is the remarkable story of First Lieutenant James L. Giangrosso:

> He led his platoon in the attack, then directed the repulse of an enemy counterattack. While withdrawing to better night positions, he noticed that two of his men were missing. Leaving his platoon in charge of his platoon sergeant, Lieutenant Giangrosso returned alone to the battlefield where he found the two men, both seriously wounded. He carried one man back under enemy fire, and then returned for the second. This man was too seriously wounded to be moved without a litter, and Lieutenant Giangrosso was advised that a litter team was unavailable until morning. Undeterred, Lieutenant Giangrosso remained in no-man's land with the wounded man all night, hiding from several enemy patrols scouring the area. Both were successfully rescued at dawn. (160)

Stories such as the above appear in nearly all the chapters and are the distinguishing feature of the book. In most chapters, Prefer also adds as a last paragraph what appears to be a synopsis of what's happening in the war's other theaters. These paragraphs are out of place and puzzling.

The following appeared at the end of Chapter II, which covered the initial landings in Lingayen Gulf:

> In Northwest Europe, the First and Third U.S. Armies were fighting to clear the ground seized by the enemy during the Battle of the Bulge. In the Seventh U.S. Army's zone, the 43d Tank Battalion of the 12th Armored Division was wiped out at Herrlisheim. On the Eastern Front, the First White Russian Front overran Warsaw. In Italy, the Fifth U.S. Army relieved British units facing the Gothic Line. The American Mars Task Force had reached to within three miles of the Burma Road. (26)

Paragraphs such as the above appeared in Chapters 2-11. The war in Europe ended May 2, which coincided with Chapter 11. There is no context, nor explanation. Prefer did not mention that perhaps he was using these chapter-ending paragraphs to measure the progress and losses of the Luzon Campaign against those prosecuted in the other theaters of World War II. Instead, the reader is left to wonder why they are even there.

This book lacks two fundamentals as a work of history: sufficient footnotes/endnotes and sources. Prefer uses endnotes but there are an alarmingly few in the book. The personal stories detailed in a previous paragraph are not cited. Specific numbers and statistics are not cited. Take into consideration Chapter 9; an example is the chapter titled "The Shimbun Group." It has numbers such as specific bomb tonnages, WIAs, and KIAs. These figures are not cited. The chapter has 22 endnotes, none of which address numbers, decisions, words, or statements that go far beyond common knowledge. Instead, one of the 22 endnotes is a detailed description of the B-24 Liberator heavy bomber's speed, range, armament, and load capacity, again without mention of where the author found his information. Another endnote explains all the casualty numbers of the 503rd Parachute Infantry Regiment. Again, where did those numbers come from? All the chapters have similar citation problems.

There is also a surprising lack of maps. Several chapters do have maps, but most do not. The maps, for the most part, have a common source: Robert Ross Smith's *Triumph In The Philippines*. All the Smith maps are excellent, but actual unit operational reports have more maps and pictures of enemy positions, terrain, and the like. When detailing battles, the author mentions movements, roads, and geography without a point of reference for the reader. A chapter is devoted to the 32nd Infantry Division's toils along the Villa Verde Trail. No pictures, no maps. It is hard to understand that battle without some visual enhancement.

The bibliography of this book is long but misleading. The secondary sources are the basis of the book, as indicated by the few endnotes provided. The primary sources are divisional histories produced by unit Public Relations Offices or individual veterans after the war. The remainder are the available but rarely used personal accounts of senior commanders such as Walter Krueger's *From Down Under To Nippon*, Robert Eichelberger's *Our Jungle Road To Tokyo*, MacArthur's *Reminiscences*, and Charles Willoughby and John Chamberlain's *MacArthur, 1941–1951*. The only one of any value is Eichelberger's book ,and he was on Luzon for only the first few weeks of the campaign. As the Sixth Army commander and responsible for this entire campaign, Krueger would be a great source for understanding the battle. His book is historically useless, containing little knowledge that would help the historian as it is majorly lacking in detail. The amount of primary source material available on this subject is massive. Each unit involved in the Luzon campaign kept extensive records of their efforts. They are readily available at the National Archives, Modern Military Records, in College Park, Maryland, and at the U.S. Military History Institute at Carlisle Barracks, Pennsylvania. Most of the records at Carlisle Barracks are also available electronically. None were consulted. They are far more accurate than the watered-down accounts of the Divisional Public Relations histories or the useless works of Krueger, MacArthur, and Willoughby.

Hopefully, Prefer's other seven titles—*The Battle for Tinian, The Conquering Ninth, Leyte 1944: The Soldier's Battle, MacArthur's New Guinea Campaign: March-August 1944, Patton's Ghost Corps: Cracking the Siegfried Line*, and *Vinegar Joe's War: Stillwell's Campaign in Burma*—are better, but I would not recommend adding Nathan Prefer's *The Luzon Campaign: MacArthur Returns* to a military history collection. At the beginning of the work, the author states that a comprehensive examination of the Luzon campaign does not yet exist. He is correct, but this book does not address that deficiency. The best work on Luzon remains Robert Ross Smith's *Triumph In The Philippines*, the U.S. Army's official history of the 1945 Philippines campaign.

About The Author

Robert Young received a Ph.D. in Military History and Modern European Studies from the C.U.N.Y. Graduate Center in 2003 and a Masters in American History from Brooklyn College in 1994. He is currently a Professor of History and Military History at American Military University. A veteran of the United States Army, he served in various leadership positions in armored and cavalry units.

Book Review: Waldemar Heckel's *Who's Who in the Age of Alexander and His Successors: From Chaironeia to Ipsos (338-301 BC)*

Quinton Carr
American Public University

Waldemar Heckel. *Who's Who in the Age of Alexander and His Successors: From Chaironeia to Ipsos (338-301 BC)*. Barnsley, S. Yorkshire, UK: Greenhill Books, 2021.

The Age of Alexander is one of the most compelling and studied periods of ancient history, and numerous works examining the political intrigues, conquests, military tactics, and lives of the leading figures have been published over the past decade. The cast of characters is huge, and the list of names can be especially difficult to follow: fifteen Alexanders, twenty-two Philips, and nine Ptolemies, for example, present a significant cognitive challenge for even the most disciplined researcher. (x) Fortunately, Waldemar Heckel has provided a solution with *Who's Who in the Age of Alexander and His Successors: From Chaironeia to Ipsos (338–301 BC).*

Heckel is a well-known Classical scholar and Alexander specialist whose published works include *The Marshals of Alexander's Empire* (1992), *The Wars of Alexander the Great* (2002), *Alexander the Great: Historical Sources in Translation* (2004), and *Who's Who in the Age of Alexander the Great: Prosopography of Alexander's Empire* (2006). While this book is derived from the 2006 title, the 2021 revisions are so extensive that it should be considered a new work rather than a revised edition, a point that is highlighted by the updated title. Differences between the new volume and the old include "names are given in their Greek forms" where known rather than the more common

 doi: 10.18278/sshj.12.1.8

Latin/Anglicized forms, and all entries are numbered (ix). There are a few exceptions to this, including Alexander (Alexandros), Philip (Philippos), and Ptolemy (Ptolemaios). There were 820 entries in the 2006 edition, which covers the period from 338 BC to Alexander's death in 323. The 1,150 entries in this volume (many of which have been updated with new content and references) include the Wars of the Successors through the Battle of Ipsos in 301.

Who's Who in the Age of Alexander and his Successors is a prosopography and is the most extensive English-language work of its kind for the period. Prosopography is "a collective biography, describing the external features of a population group that the researcher has determined has something in common (profession, social origins, geographic origins, etc…)"[1] Heckel assumes that the reader understands this definition and does not explicitly mention it in the 2006 or 2021 volumes. In effect, it is an encyclopedia of mini-biographies of people belonging to a certain period, along with other supporting material. *Who's Who* includes an excellent "Foreword" by Peter G. Tsouras, Lt. Col., USAR (ret.), "Preface," "Abbreviations" (vital for deciphering the complex references), "The Biographies," "Anonymous Individuals," "Stemmata" (family trees), extensive "Bibliography," and "Concordance," an index of Latinized names from the 2006 work that links to their Greek form entries.

Peter Tsouras's Foreword describes the formation of Heckel's plan for this "significant clarification and expansion" of the 2006 work (ix). Tsouras explains the major differences between the two books and the project's magnitude; his words segue nicely into Heckel's Preface.

The 1,150 biographical entries range from a single paragraph for minor personalities to several pages for the most famous individuals: more than three for Philip, five for Perdikkas, eight-plus each for Alexander, Antigonos, and Ptolemy. Each entry presents the individual's name in Greek, discusses their lineage and early life, summarizes their actions during the period, and concludes with a discussion of their character and significance. All are heavily referenced with in-text citations, which is valuable and navigable for the academic researcher but awkward for casual readability. This underscores that Heckel's target audience is the professional Classicist or historian rather than the general reader. A good example is the first two sentences of the entry for Antigonos "the One-Eyed":

> **115 ANTIGONOS**
>
> (Ἀντίγονος). Nicknamed "the One-Eyed": Monophthalmos, Heterophthalmos or Kyklops (Ael. VH 12.43; Plut. Ser. 1.8; Pliny, HN 35.90; Polyb. 5.67.6; cf. Plut. Mor. 633c). Macedonian (Tataki 245 no. 169). A. was born not long after 386/5 (Hieronymos ap. [Lucian], Macrob. 11 = FGrH 154 F8; App. Syr. 55.279; cf. Plut. Demetr. 19.4, who says that in 306 he was "a little short of 80 years old"; Porphyry of Tyre, FGrH

> 260 F32 says he was 86 at the time of his death!), the son of a certain Philip (Arr. 1.29.3; Syll3 278, 5; Strabo 12.4.7 C565; Justin 13.4.14; Ael. VH 12.43; Plut. Demetr. 2.1; Hieronymos ap. [Lucian], Macrob. 11 = FGrH 154 F8). (56)

Once the obstacle course of abbreviations and referential asides has been navigated, the content presented is worth the perseverance. Heckel has included an impressive amount of information in each summary, and the value of the references becomes clear when the researcher is looking for more context and detail.

Heckel sometimes digresses into academic controversies in the biographies, especially when discussing perceived character traits. When addressing speculation about Alexander's sexuality, he writes: "Attempts to see Olympias as the cause of A.'s alleged sexual insecurities have met with only limited success, but the modern scholar will interpret these personal details in accordance with his or her own predilections (and social milieu), and the reader will believe whatever he or she chooses, often with little regard for cultural context." (25) These excursions sometimes interrupt the flow of the narrative but give the reader a broader perspective on the differing scholarly views of a given topic.

Further demonstrating his penchant for thoroughness, Heckel has included an "Anonymous Individuals" section of 105 entries. The majority are women—wives, daughters, sisters—but also features male minor rulers, officials, and relatives of named individuals. Here is an interesting example:

> **A16** An unnamed Illyrian queen, killed in single combat by the young princess Kynnane (no. 634), the daughter of Philip II and Audata-Eurydike (Polyaenus 8.60). Berve II 229 dates her death to c..344/3, i.e., before Alexander's accession. This seems somewhat early, since Kynnane was born c.357, and there is no reason to equate every skirmish with Illyrians with the famous battle of 344/3! But, since Polyaenus mentions Kynnane's exploits before her marriage to Amyntas son of Perdikkas (no. 76), the battle in question probably occurred late in Philip's reign. (501)

The "Stemmata" is a collection of genealogical diagrams (family trees) of the Macedonian Royal House (the Argeads), Molossion Royal House, the Achaimenids, Pharnabazos, Iolaos, Antigonos, Pamenion, Agathokles, Ptolemy, Seleukos, and several others. The diagrams provide the reader with clear visual aids that further assist in sorting through the tangle of names and relationships of the major players of the period. (501–502)

Who's Who in the Age of Alexander and his Successors is a monumental reference work that showcases Heckel's decades of meticulous scholarship. It is an invaluable tool as a sourcebook and navigation aid for the period. Academics will find it a versatile companion to Arrian, Diodorus, Justin, Plutarch, Quintus Curtius Rufus, and the most important modern sources. While non-professionals may find the style of in-text citations and Heckel's occasional digressions challenging, those with a strong interest in Alexander and the Successors will find the work a treasure trove of information and well worth the effort.

Endnote

1 Carlier, Myriam, Dumolyn, Jan, and Koenraad Verboven. "A Short Manual to the Art of Prosopography" in Katharine S. B. Keats-Rohan, ed., *Prosopography Approaches and Applications: A Handbook*, Prosopographica et Genealogica 13 (Oxford: Unit for Prosopographical Research, Linacre College, Univ. of Oxford, 2007), 39.

About the Author

Quinton Carr is pursuing a master's degree in history with a concentration in Ancient and Classical History at American Military University. A native Vermonter, he spent eight years on active duty with the U.S. Navy, earned a BA in History from the University of Southern Maine, and an MBA from Franklin Pierce University. Quinton currently works for NewsBank, Inc., as Director of Publisher Support in their Chester, Vermont location.

Reimagining the Islandmagee Witches: An Interactive Exhibit at Carrickfergus Museum

Gilda Bour

Independent Historian

The last witch trial in Ireland was brought alive this past autumn as part of an exhibit at the Carrickfergus Museum in Northern Ireland. While it was set up in a single room, the small space delivered quite a powerful historical experience with a rich assortment of mediums ranging from antiquities to virtual reality, videos, and a comic book detailing and illustrating the event. The options gave visitors the ability to learn about and explore the Islandmagee witch trials in many different formats, bringing them closer to what occurred over two hundred years ago. The exhibit space appeared larger than it was and provided visitors the ability to immerse themselves in a choose-your-own-adventure type of experience.

The exhibit tells the story of what is widely believed to be the last trials for witchcraft in Ireland in 1711, following events on the peninsula of Islandmagee, County Antrim. In February of 1711, a young woman, Mary Dunbar, traveled with her cousin to Islandmagee. She came following the death of her elderly aunt, Ann Haltridge, the widow of a Presbyterian minister, who had died suddenly. Before her death, it was believed that Ann had suffered for months from demonic attacks. Soon after arriving in the area, Mary Dunbar herself began to exhibit symptoms of demonic attacks. These included vomiting household items, levitation, and convulsions.

Mere days after her arrival, Mary began accusing multiple women of

A poster advertising the museum exhibit, hanging near Carrifckfergus Castle. [Carrickfergus, Antrim, Northern Ireland. Photo by Gilda Bour.]

doi: 10.18278/sshj.12.1.9

witchcraft. The women were from Islandmagee and surrounding areas. The inhabitants in and around the peninsula feared the Devil more than most and, as such, tended to attribute unknown ailments and unexplained phenomena to the work of the Devil.[1] Mary was considered a credible and believable witness as a comely, smart, and educated woman of 18.

At the beginning of March, the first woman Mary Dunbar accused as her tormentor was Janet Carson. Then a few days later, Mary named a woman named Catherine. All women in Islandmagee named Catherine were paraded before Mary. Mary was calm in front of the women, but fell into fits and screamed in the presence of Catherine McCalmond. Additional women were named and arrest warrants were issued for Janet Carson, Catherin McClamond, Elizabeth Sellor, and Janet Liston.[2] The women were brought before Mary Dunbar, and she fell into fits and begged for them to be removed from her presence. This was seen as proof of her torment by witchcraft, and the women were arrested and sent to Carrickfergus, where they were imprisoned.

Mary Dunbar continued to have fits and provided descriptions and names of her tormentors. Janet Latimer, Margaret Mitchell, and Janet Main were arrested following the arrest of the first four women. These would not be the last women arrested, as Mary Dunbar continued to exhibit demonic fits, and under questioning and testimony described more women. Janet Millar, a woman smoking tobacco and walking down the road, fit the description Mary had provided of a tormentor, and when confronted she swore and was arrested. Janet Main was interrogated about her faith and was surly with her answers. When she was put before Mary Dunbar, Mary again fell into fits and screamed for Janet to be removed. This again provided "proof" and Janet Millar was formally arrested. All of the women arrested did not fit the standards of beauty and behavior expected, and they shared traits from arthritis to deformities, surly attitudes, and outspoken manners.

At the end of March, Mary Dunbar prepared to travel to the trial of the eight women she had accused. During the journey, she experienced fits and fell from her horse. At the trial, she was unable to lead the prosecution, as she began to fit as soon as the trial started. Many witnesses gave testimony for the prosecution. The jury and judges summed up and delivered a verdict of "guilty of exercising of witchcraft on the body of Mary Dunbar."[3] In accordance under the 1586 Irish Witchcraft Act, the women were found guilty on March 31, 1711, and sentenced to be pilloried four times and serve one year in prison.[4]

After the trial, Mary Dunbar accused two women and a man of tormenting her during the trial. On her journey back to her home in the following week, Mary experienced more fits and vomited pins. She described her tormentor as a man, and the father of Elizabeth Sellor and husband of Janet Liston was identified and arrested.[5] He was brought before Mary, but she refused to identify him, claiming later

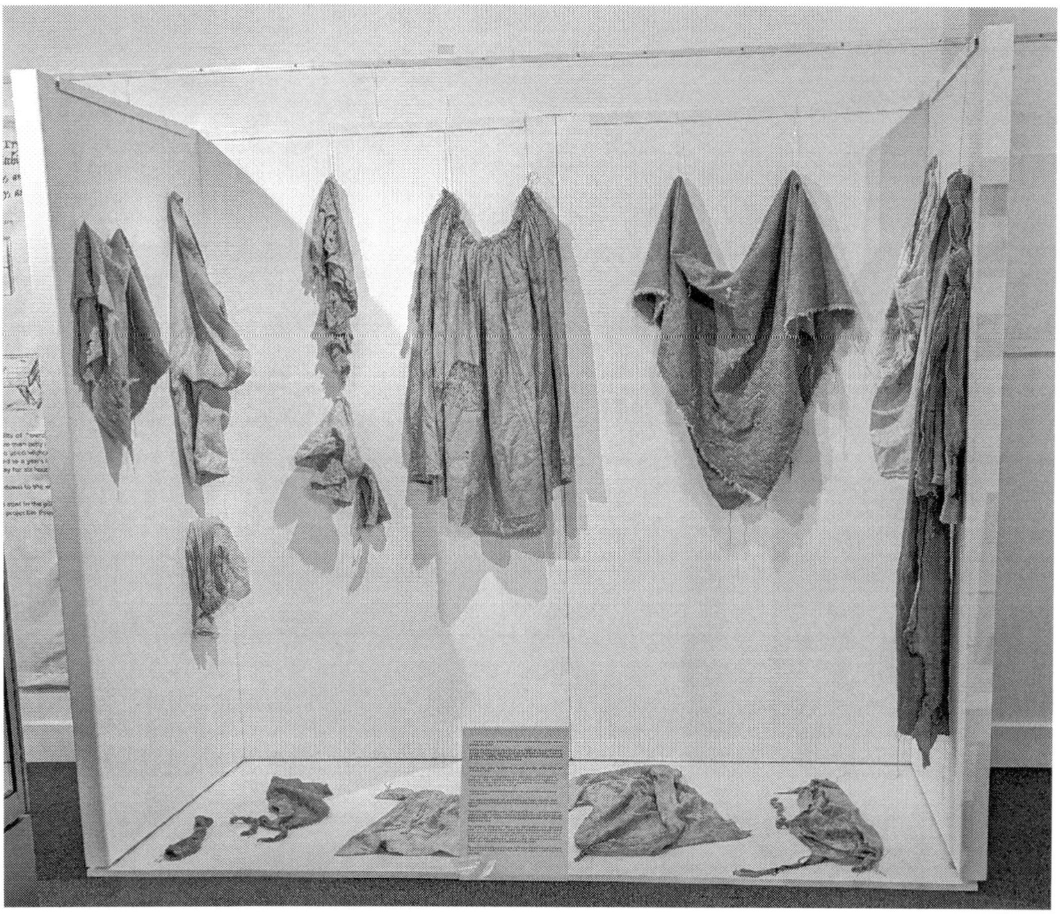

A collection of linen and woolen cloth, that ranges from unblemished to dirtied and burned. [Carrickfergus Museum. Photo by Gilda Bour.]"

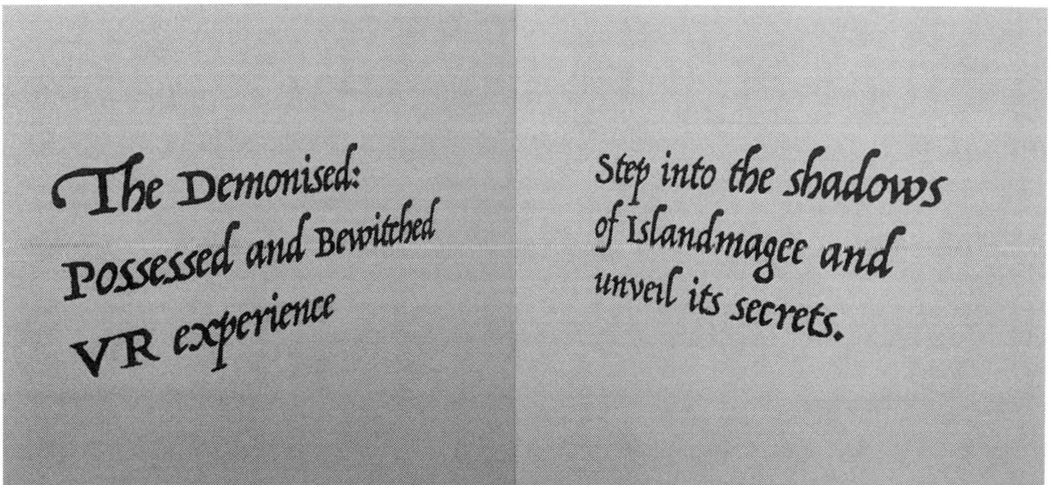

Display area title showcasing the VR Experience Demonised: Possessed and Bewitched. [Carrickfergus Museum. Photo by Gilda Bour.]

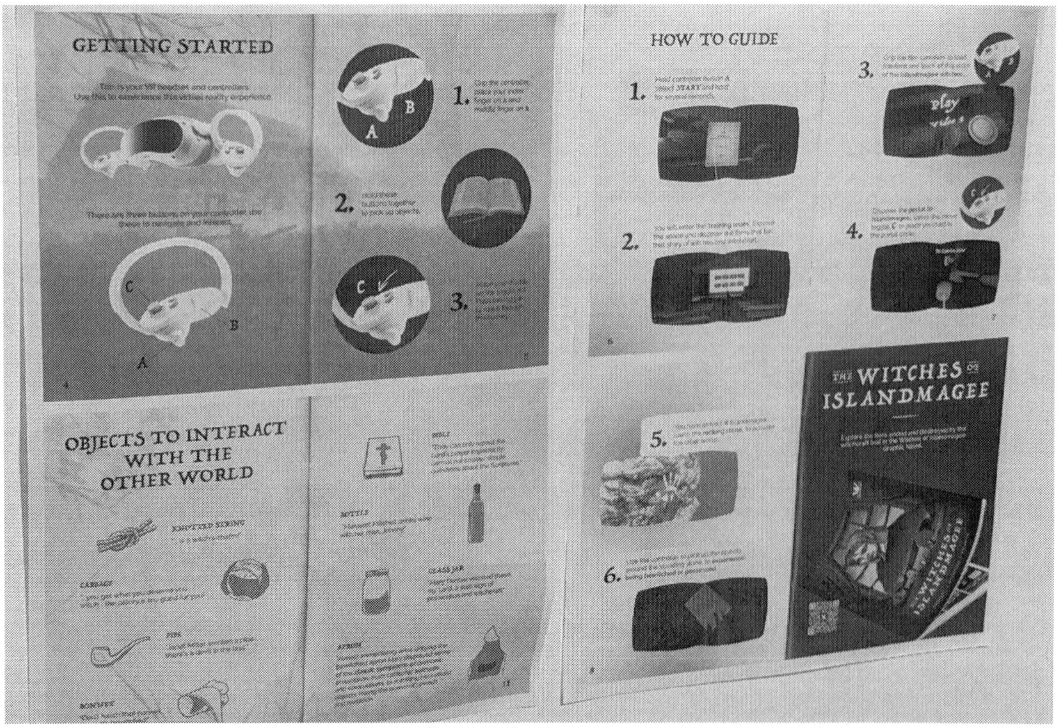

Instructions for the virtual reality experience of the exhibit.
[Carrickfergus Museum. Photo by Gilda Bour.]

Video displays and comic book portrayal of Islandmagee Witch Trial.
[Carrickfergus Museum. Photo by Gilda Bour.]

she feared for her safety. William Sellor was known to locals for his temper and drinking. William Sellor's trial was to follow after his arrest, but his accuser, Mary, died before his trial took place. Due to this, his charge was converted to a capital offense, of which he was convicted on September 1711.[6] It is believed that he was put to death but due to the lack of records, this is unconfirmed. Mary Dunbar's death concluded the largest and last witch trials in Ireland.

Upon entering the room the exhibit is housed in, visitors are presented with a type of choose-your-own-adventure exhibit experience. They can meander around the room and look at artifacts from the early 18th and 19th centuries such as an old stool, shackles, and old books that contain the details of the trial, the people, and an eyewitness account. A wall with clothing presented as unblemished and then torn, dirty, and burned shared a unique visual perspective of the clothing of one accused. The material showed how the clothing of an accused townsperson might change as they went from that of a standard townsperson to a jailed person, unable to wash or perform basic cleaning tasks and even worse, to a dismal fate. The use of this medium was contemplative and creative, causing one to pause and consider how the experiences of a person are also shared through the clothing one wears. Large posters around the room presented the witch trial history with quotes, illustrations, and summaries that gave vital information as to what occurred, and in a pleasing and easy-to-digest manner.

The most robust parts of the exhibit were the unexpected and new methods of presentation, which included mediums such as a comic book, tablets to watch videos, and virtual reality to make the experience come alive in an unexpectedly realistic fashion via an

Letter with accounting of conversations with trial judges.
[Carrickfergus Museum. Photo by Gilda Bour.]

A narrative book detailing the events of Islandmagee in 1711.
[Carrickfergus Museum. Photo by Gilda Bour.]”

A first edition book copy of Professor Andrew Sneddon's, Possessed by The Devil: The Real History of the Islandmagee Witches and Ireland's Only Mass Witchcraft Trial.
[Carrickfergus Museum. Photo by Gilda Bour.]

interactive video game. The comic book took the witch trials into an illustrated graphic format and provided a pleasing albeit harrowing reading experience. The tablets included eight videos that allowed the user to watch the events unfold and witness history told through an artful cinematic lens. The virtual reality option provided a fantastic interactive experience that utilizes cutting-edge technology to reach a wider audience and bring history closer. With this method, visitors were transported into Mary's experience to learn more about what it might have been like to be possessed—or, alternatively, choose in-

Shackles that would have been used for the accused in the trials."
[Carrickfergus Museum. Photo by Gilda Bour.]"

Witch stones, also known as elf or hag stones, were used common protections against harmful magic in the Ulster area. [Carrickfergus Museum. Photo by Gilda Bour.]

stead to be one of the accused. As Mary, one is presented with the question of whether one would accuse someone, or not. Visitors get to choose how they would react when confronted with a similar moral dilemma. Alternatively, if visitors choose to be one of the accused, they are presented with the experience of potentially facing a penalty of death. The setting of the video game showed the town as it would have been in the early part of the 18th century, in a darkened fashion that enhanced the experience to provide a feeling of risk, like those involved in the incident.

While the exhibit ran from September to November 11, 2023, this author hopes the exhibit will be extended, become permanent, or be brought back each autumn. This was one of the drawbacks, unfortunately, that such a powerful, immersive exhibit experience was limited to a short run and if not already in the area, it was easily missed. The museum does make up for this with its other exhibits. It has much to offer on local and Irish history, and with no cost to attend, it should be visited and explored. There are permanent exhibits and touring exhibits that range from medieval to recent periods.

The Carrickfergus Museum is open Monday – Saturday, 10 a.m. – 4 p.m., and admission is free. The museum is conveniently located in the Carrickfergus Museum & Civic Centre, and the contact information is:

Carrickfergus Museum
Carrickfergus Museum and Civic Centre
11 Antrim Street
Carrickfergus
BT38 7DG
Telephone: 028 9335 8241
Email: carrickfergusmuseums@midandeastantrim.gov.uk

Notes

1 Andrew Sneddon, *Possessed by The Devil: The Real History of the Islandmagee Witches and Ireland's Only Mass Witchcraft Trial* (Dublin, Ireland: The History Press Ireland, 2013), 38.

2 Ibid, 8.

3 Ibid, 151.

4 Andrew Sneddon, *Witchcraft and Magic in Ireland* (Basingstoke, Hampshire, England: Palgrave MacMillan, 2015), 84.

5 Ibid, 89.

6 Victoria McCollum and Andrew Sneddon, "About the Witches of Islandmagee Project," Witches of Islandmagee, February 7, 2024, https://w1711.org/history/.

Bibliography

McCollum, Victoria, and Andrew Sneddon. "About the Witches of Islandmagee Project." Witches of Islandmagee, February 7, 2024. https://w1711.org/history/.

Sneddon, Andrew. *Possessed by The Devil: The Real History of the Islandmagee Witches and Ireland's Only Mass Witchcraft Trial*. Dublin, Ireland: The History Press Ireland, 2013.

Sneddon, Andrew. *Witchcraft and Magic in Ireland*. Basingstoke, Hampshire, England: Palgrave MacMillan, 2015.

This publication is available open access at:
http://www.ipsonet.org/publications/open-access

Thanks to the generosity of the American Public University System

American Public University APU

American Military University AMU

Conferences

Policy Studies Organization hosts numerous conferences, including the Middle East Dialogue, Space Education and Strategic Applications, International Criminology Conference, Dupont Summit on Science, Technology and Environmental Policy, World Conference on Fraternalism, Freemasonry and History, AI – The Future of Education: Disruptive Teaching and Learning Models, Sport Management and Esport Conference, and the Internet Policy & Politics Conference. Recordings of these talks are available in the PSO Video Library.

Yearbook

The Policy Yearbook contains a detailed international listing of policy scholars with contact information, fields of specialization, research references, and an individual scholar's statements of research interests.

Curriculum Project

The Policy Studies Organization aims to provide resources for educators, policy makers, and community members, to promote the discussion and study of the various policies that affect our local and global society. Our curriculum project organizes PSO articles and other media by easily serachable themes.

For more information on these projects, access videos of past talks, and upcoming events, please visit us at:

ipsonet.org

Related Titles from Westphalia Press

The Limits of Moderation: Jimmy Carter and the Ironies of American Liberalism by Leo P. Ribuffo

The Limits of Moderation: Jimmy Carter and the Ironies of American Liberalism is not a finished product. And yet, even in this unfinished stage, this book is a close and careful history of a short yet transformative period in American political history, when big changes were afoot.

The Zelensky Method by Grant Farred

Locating Russian's war within a global context, The Zelensky Method is unsparing in its critique of those nations, who have refused to condemn Russia's invasion and are doing everything they can to prevent economic sanctions from being imposed on the Kremlin.

Sinking into the Honey Trap: The Case of the Israeli-Palestinian Conflict by Daniel Bar-Tal, Barbara Doron, Translator

Sinking into the Honey Trap by Daniel Bar-Tal discusses how politics led Israel to advancing the occupation, and of the deterioration of democracy and morality that accelerates the growth of an authoritarian regime with nationalism and religiosity.

Notes From Flyover Country: An Atypical Life & Career by Max J. Skidmore

In this remarkable book, Skidmore discusses his "atypical life and career," and includes work from his long life in academe. Essays deal with the principles and creation of constitutions, anti-government attitudes, the influence of language usage on politics, and church-state relations.

The Athenian Year Primer: Attic Time-Reckoning and the Julian Calendar by Christopher Planeaux

The ability to translate ancient Athenian calendar references into precise Julian-Gregorian dates will not only assist Ancient Historians and Classicists to date numerous historical events with much greater accuracy but also aid epigraphists in the restorations of numerous Attic inscriptions.

Siddhartha: Life of the Buddha by David L. Phillips, contributions by Venerable Sitagu Sayadaw

Siddhartha: Life of the Buddha is an illustrated story for adults and children about the Buddha's birth, enlightenment and work for social justice. It includes illustrations from Pagan, Burma which are provided by Rev. Sitagu Sayadaw.

Growing Inequality: Bridging Complex Systems, Population Health, and Health Disparities Editors: George A. Kaplan, Ana V. Diez Roux, Carl P. Simon, and Sandro Galea

Why is America's health is poorer than the health of other wealthy countries and why health inequities persist despite our efforts? In this book, researchers report on groundbreaking insights to simulate how these determinants come together to produce levels of population health and disparities and test new solutions.

Issues in Maritime Cyber Security Edited by Dr. Joe DiRenzo III, Dr. Nicole K. Drumhiller, and Dr. Fred S. Roberts

The complexity of making MTS safe from cyber attack is daunting and the need for all stakeholders in both government (at all levels) and private industry to be involved in cyber security is more significant than ever as the use of the MTS continues to grow.

Female Emancipation and Masonic Membership: An Essential Collection By Guillermo De Los Reyes Heredia

Female Emancipation and Masonic Membership: An Essential Combination is a collection of essays on Freemasonry and gender that promotes a transatlantic discussion of the study of the history of women and Freemasonry and their contribution in different countries.

Bunker Diplomacy: An Arab-American in the U.S. Foreign Service by Nabeel Khoury

After twenty-five years in the Foreign Service, Dr. Nabeel A. Khoury retired from the U.S. Department of State in 2013 with the rank of Minister Counselor. In his last overseas posting, Khoury served as deputy chief of mission at the U.S. embassy in Yemen (2004-2007).

Managing Challenges for the Flint Water Crisis Edited by Toyna E. Thornton, Andrew D. Williams, Katherine M. Simon, Jennifer F. Sklarew

This edited volume examines several public management and intergovernmental failures, with particular attention on social, political, and financial impacts. Understanding disaster meaning, even causality, is essential to the problem-solving process.

User-Centric Design by Dr. Diane Stottlemyer

User-centric strategy can improve by using tools to manage performance using specific techniques. User-centric design is based on and centered around the users. They are an essential part of the design process and should have a say in what they want and need from the application based on behavior and performance.

Masonic Myths and Legends by Pierre Mollier

Freemasonry is one of the few organizations whose teaching method is still based on symbols. It presents these symbols by inserting them into legends that are told to its members in initiation ceremonies. But its history itself has also given rise to a whole mythology.

How the Rampant Proliferation of Disinformation has Become the New Pandemic by Max Joseph Skidmore Jr.

This work examines the causes of the overwhelming tidal wave of fake news, misinformation, disinformation, and propaganda, and the increase in information illiteracy and mistrust in higher education and traditional, vetted news outlets that make fact-checking a priority

Anti-Poverty Measures in America: Scientism and Other Obstacles
Editors, Max J. Skidmore and Biko Koenig

Anti-Poverty Measures in America brings together a remarkable collection of essays dealing with the inhibiting effects of scientism, an over-dependence on scientific methodology that is prevalent in the social sciences, and other obstacles to anti-poverty legislation.

Geopolitics of Outer Space: Global Security and Development
by Ilayda Aydin

A desire for increased security and rapid development is driving nation-states to engage in an intensifying competition for the unique assets of space. This book analyses the Chinese-American space discourse from the lenses of international relations theory, history and political psychology to explore these questions.

Contests of Initiative: Countering China's Gray Zone Strategy in the East and South China Seas
by Dr. Raymond Kuo

China is engaged in a widespread assertion of sovereignty in the South and East China Seas. It employs a "gray zone" strategy: using coercive but sub-conventional military power to drive off challengers and prevent escalation, while simultaneously seizing territory and asserting maritime control.

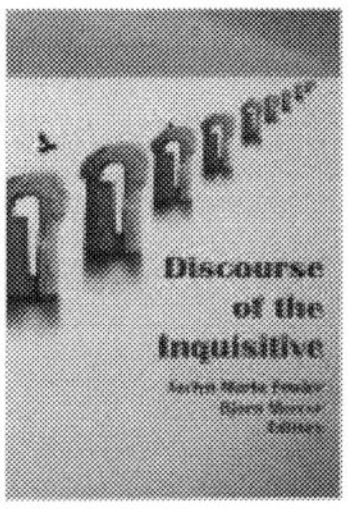

Discourse of the Inquisitive
Editors: Jaclyn Maria Fowler and Bjorn Mercer

Good communication skills are necessary for articulating learning, especially in online classrooms. It is often through writing that learners demonstrate their ability to analyze and synthesize the new concepts presented in the classroom.

westphaliapress.org

Made in the USA
Monee, IL
08 November 2026

68002050R00072